An Introduction to
RISK ANALYSIS

Books by Robert E. Megill

An Introduction to
Exploration Economics, 1971

How to Be A Productive Employee, 1973

Published by
Petroleum Publishing Company
Tulsa, OK 74101

An Introduction to

RISK
ANALYSIS

Robert E. Megill

Petroleum Publishing Company
1977

This book is dedicated to
my beloved wife, Margaret.
She has always been at my side in any endeavor—
constantly supportive and seldom critical.
I owe her more than
she will ever know.

Contents

ACKNOWLEDGMENTS

Special thanks are due to Messrs. R. I. Swanson, C. R. Clark and J. E. Smitherman. Each read early drafts of the manuscript and gave important criticism. Their efforts contributed to a better finished product.

To G. Rogge Marsh, who has reviewed carefully all of my published works to date, go my deepest thanks and appreciation. He not only provided important criticism, but contributed ideas which were incorporated in the final draft. Each of the prior books owes much to his insights and candid questions. This book also benefited from his questioning and I am sincerely grateful.

To Margaret Richards, my thanks for her valuable help in proofing the final manuscript for mathematical errors.

To my wife, Margaret, my thanks for putting up with me during the three years of research and writing and for typing each draft. It could not have been done without her.

Preface

My first published book, a success in a narrow field, was a simplification of established economic concepts; my second a statement of opinion about a personal work ethic and personal productivity.

This book presents a little of each. Part is simplification of established concepts and part is opinion. Presentations of opinion carry their own burdens, especially when they refer to an area, such as risk analysis, where many opinions are possible. Any writing on risk analysis can never be other than part opinion and can also never be the "end-all" to a problem with innumerable solutions.

It is hoped however that both the mathematical simplifications and opinions on risk analysis will provide a beginning—an introduction to a very complex subject.

Risk analysis is becoming a bonafide segment of all modern decision-making, economic evaluation and forecasting. If this treatise can help to broaden the base of understanding and increase the number of persons with a beginning comprehension of the fundamentals of risk analysis, it will have served its purpose. In this book you will find many useful references at the end of each chapter. These have been carefully selected for their ease of reading. The more complex writings in the area of risk analysis have been omitted. However, as you read further of risk analysis you will find other references which will increase your knowledge. It is my fond hope that what you read in this book will spur you on to further reading.

The words which follow were not recorded for the expert. They are a personal effort to make simple a few of the basic fundamentals in the analysis of risk. Every chapter has been constructed with the underlying purpose of providing a truly introductory explanation to an important method or concept in risk analysis. Only you can judge whether this goal was achieved.

How to Use This Book

"An Introduction to Risk Analysis" is conceived as a companion volume and sequel to my earlier book entitled "An Introduction to Exploration Economics." It is in essence a major expansion of Chapters 8 and 9 of "An Introduction to Exploration Economics," which dealt but briefly with an analysis of risk. The book will borrow a few illustrations from "Exploration Economics." It will not, however, cover the economic analysis of prospects or plays which the earlier book reviewed.

In keeping with its predecessor, this book assumes that the reader must start from the very beginning on all concepts. It is divided into two parts. Part I deals with simple mathematical concepts which will be helpful in understanding Part II. If you already have a strong statistical background you may wish to skip directly to Part II, which deals with the important concepts and methods in risk analysis. If not, you may wish to review specific chapters prior to beginning the second part. Portions of the book can be read in an isolated manner; i.e. a good many of the chapters stand on their own. An example is the chapter on competitive bidding. Since it is primarily a review of published literature, it can be read without considering most of the other chapters. Even here, however, it is of value to understand the importance of lognormality in oil and gas field-size relationships.

Throughout the book the plus as well as the minus aspects are reviewed for each concept. Anyone wanting to utilize the basic concepts of risk analysis will still find some problems in acceptance. The fear that a few geologists have of the quantification of uncertainty is akin to the early fears about the use of the computer in mapping. This fear stems largely from misinformation and lack of knowledge of both the benefits and limitations of risk analysis. The difficulty in comprehending a new concept always stands in the way of its acceptance. Not everyone appreciates the value of an analysis of risk. You need to understand the psychological problems associated with the hangups potential users of your work may have.

In my opinion much of the fear about using the techniques of risk analysis is the fault of the "teacher" and not the "student." Many are willing to learn, but in all too many cases the teacher has failed to show in easily understandable terms the beneficial self-interest of risk analysis.

The following paragraphs briefly describe each chapter. From the description you may determine for Part I just what reading is necessary for you or of special interest, or whether you can go immediately to Part II.

In Chapter One the ground-work is laid to show the value of a histogram in sorting masses of data. A cumulative frequency distribution is introduced.

Chapter 2 dwells on the binomial and normal distributions. A penny tossing example is used to show important characteristics of a binomial distribution. If you already understand binomial distributions (two outcome distributions) skip to the next chapter.

In Chapter 3 the characteristics of a distribution are defined and illustrated. You need to understand thoroughly the mean and standard deviation. They are the two key characteristics most frequently used for any distribution.

Lognormal distributions are the subject of Chapter 4. This chapter is important background for later discussions on field size distributions and competitive bidding. In Chapter 4 you will learn why calculations involving multiplication yield lognormal distributions.

Chapter 5 deals with histograms and frequency distributions and their shapes when converted to cumulative frequency distributions. This chapter is of particular value for those who have trouble mentally going from the shape of a distribution to its cumulative frequency.

Chapters 6 and 7 prepare the reader for a discussion of Gambler's Ruin. They provide background data on permutations, combinations and binomial probability.

Chapter 8 is the beginning of Part II. It discusses Gambler's Ruin; it explains what is meant by a "normal run of bad luck."

Chapter 9 discusses the need for analyses of opinion; it presents the pro and con aspects of subjective probability.

Because triangular distributions (3 points) are so popular, Chapters 10 and 11 illustrate certain facts about this tool.

In Chapter 12 an example of the use of opinion analysis is given. Chapter 13 illustrates how types of uncertainty can be quantified for later use in a computer model.

Basin assessment is reviewed in Chapter 14. The important role of attainable potential and field size distributions is reviewed.

Competitive bidding is discussed in Chapter 15; and the final chapter, 16, lists the fundamentals of risk analysis. It gives a brief review of steps in any analysis of risk.

Appendix A contains an expanded mathematical explanation of why the three values of maximum, minimum and most likely for a triangular distribution completely describe the probabilities shown by its cumulative frequency distributions.

Appendix B illustrates two methods for quickly determining the mean of a cumulative frequency distribution.

Appendix C contains several tables of individual and cumulative binomial probabilities.

In your future reading and work in the analysis of risk, perhaps a few ideas from this effort will help your explanations of the fundamentals.

Part I
MATHEMATICAL CONCEPTS

1 Managing Lots of Data

In business and professional fields men and women are daily confronted with masses of data. Often these masses of data are more than the human mind can grasp or comprehend. No profession escapes this dilemma. Faced with this problem what do we do? How does one manage lots of data?

In the first part of this book, we are going to explore mathematical concepts which help to manage data and build understandings of what the data mean. We begin the search with two fundamental but simple principles.

1. We manage masses of data by simplifying and searching for meaning.
2. We do this by—
   sorting
   classifying
   analyzing.

The Act of Sorting

We sort data for meaning by searching for and finding a common base, characteristic or concept. We may sort by definition or objective. For instance, how many bits of the data fit a given characteristic we wish to measure? We can sort by time sequences, age, dimension, color or many other characteristics.

We often have pre-conceived ideas about data and its separation. Thus, separations (or even the data selected) may not always be independent of the thoughts or theories for which they were collected. The *a priori* influence in data selection presents a greater danger on unfamiliar ground; yet ignoring the personal bias or prejudice can trip you up even in an area of experience.

The establishment of information systems has increased the explorationists' ability to obtain sorted data quickly. A well data system usually stores data on a hierarchial system. Such a system permits rapid sorting and retrieval of precisely defined segments of data.

A Well Data System

In the United States the American Petroleum Institute (API) has established a unique number for each well drilled. To this number, in a well data system, is "attached" all of the pertinent data about a well. A well data system usually sorts wells first by their spot on the globe. A hierarchial system might be:

Level of Separation	Description of Separation
1	State
2	County
3	Township and range
4	Section
5	Location within the section

These data would be the first to be attached or added to the data bank of a particular well. Other data are added depending upon frequency of usage and storage capacity of the system. Geologic data, such as formations tested, formations penetrated, logs run, etc., can all be recorded for a particular well. Any geologist working with a data bank gets the benefit of much sorting by those who stored the data in the system originally.

Other Systems

The geological profession is not the only one to value data sorting. A doctor can establish a data system for his patients and "attach" records to the name of each patient. The Federal Government and many commercial institutions attach data to social security numbers. A few cities store data about property. Rather than a complex property description, the data can be attached to the geographic location (latitude-longitude) defined as the center of the property. Utility companies set up data banks on their electric lines with the telephone pole as the equivalent of an oil or gas well to which data are attached.

What would you record about a telephone pole? You might be surprised at the facts available for storing which are useful to a utility company. Here are some for a start:

1. Geographic location
2. Distance and direction to preceeding and following pole
3. Number of cross bars (arms)
4. Wires per cross bar
5. Voltage carried by wires
6. Wire size or type
7. Age of pole
8. Type of pole (aluminum, creosoted, etc.)
9. Number of transformers per pole.

Can you think of other facts?

The point of our digression is this—when you have lots of facts you can begin to get meaning by carefully sorting these facts.

Classifying

Perhaps sorting and classifying have the same meaning to you. As used herein, classification is an interim step between sorting and analyzing. A rough sort of data can still leave much classifying to be done. Sorting the wildcat wells may leave much additional sifting if you are looking for only those wildcats with gas shows in a specific formation. Even when you have separated these wells you may wish to make further refinements in your selections, such as:

gas shows in sands thicker than 15 ft,
gas shows confirmed by sidewall core,
and so forth.

Classification can be thought of as grouping things already sorted. Again we group for meaning and simplification. Once data have been sorted to give the ability to retrieve selected portions, we can perform other calculations which help classify the data.

Grouping data by percentage is an example of classification. If you have sorted out all the wells that were wildcats and which were offshore, you might want to classify them by some other characteristic such as depth. If you use a common class interval (such as every 2,500 ft) you arrive at the basis for a frequency histogram.

In Table 1.1 a set of data is provided which will be used for several figures. The data concern a sorting and classification of wells, drilled offshore, in a specific year (1970) in a given area.

Analyzing What is Classified—A Distribution

The data in Table 1.1 show how 365 wells can be grouped into depth intervals of 2,500 ft. These data can be plotted two ways, as shown in Figs. 1.1 and 1.2.

In Fig. 1.1 the wells are plotted by the number occurring in each class or depth interval. The most common depth interval is from 7,500 to 10,000 ft with 100 of the 365 wells bottomed in this interval.

In Fig. 1.2 the vertical scale is not the number of wells but their

TABLE 1.1

OFFSHORE WELLS
AREA A—1970

1	2	3	4	5
Total Depth	No. of		Percent	
(Class Interval)	Wells	Of Total	Cumulative*	
			Shallower Than	Deeper Than
0–2,500	20	5	5	100
2,501–5,000	40	11	16	95
5,001–7,500	60	16	32	84
7,501–10,000	100	27	59	68
10,001–12,500	80	22	81	41
12,501–15,000	40	11	92	19
15,001–17,500	20	5	97	8
17,501–20,000	5	3	100	3
Total	365	100		

*Cumulative percentages refer to the high side of the class interval in column 4, and low side in column 5.

frequency, expressed as a percent of the total wells. Fig. 1.2 is a plot of the data shown in column 3 of Table 1.1. Note the change in appearance from Fig. 1.1. The shape of the graph in Fig. 1.2 is flat relative to the first figure. The flatness is a function of the vertical scale. It could have the same shape as Fig. 1.1 by expanding the vertical scale. Fig. 1.2 tells us that even though 100 wells were drilled to depths between 7,500 and 10,000 these 100 wells represent only 27% of the total wells drilled.

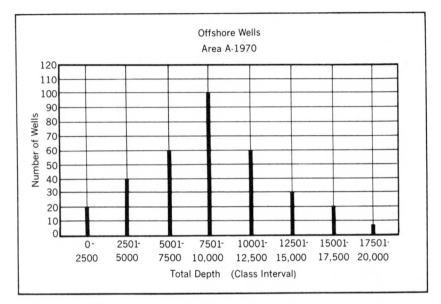

FIG. 1.1

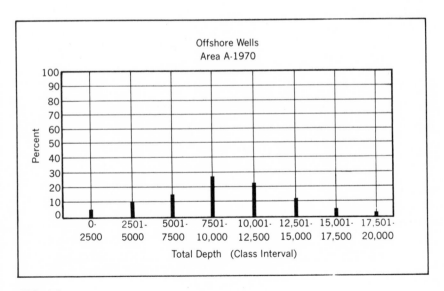

FIG. 1.2

Cumulative Frequency Distributions

The percentage graph is very useful in comparing one distribution to another. It also allows us to compile a second graph which has widespread use in risk analysis. This use is illustrated by Figs. 1.3 and 1.4 which were constructed from the data in column 4 in Table 1.1. Column 4 represents the cumulative frequency of the wells drilled. In Fig. 1.3 it is plotted as "stairsteps" rising to 100%. In Fig. 1.4, the cumulative percentage points appear as individual points connected by a smooth line.

Both figures, however, tell the same story as the fourth column in Table 1.1. They say, for example, that 81% of all of the 365 wells were in the 10,001–12,500-ft depth interval or shallower; or to put it another way, only 19% of the wells were deeper than 12,500 ft. You now see the usefulness of the cumulative percentage distribution over the numbers alone. They tell instantly how many wells were deeper or shallower than a certain percentage. We will find this graph extremely useful in the chapters ahead.

Sometimes the data on cumulative frequencies are plotted slightly differently shifting the emphasis to deeper or greater than. To graph the data this way we use the percentages from column 5 in Table 1.1. They are plotted on Fig. 1.5. This figure shows the same basic data

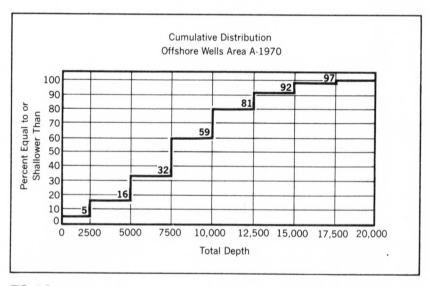

FIG. 1.3

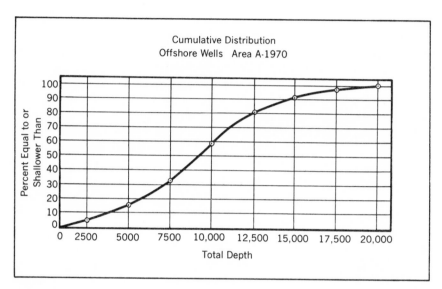

FIG. 1.4

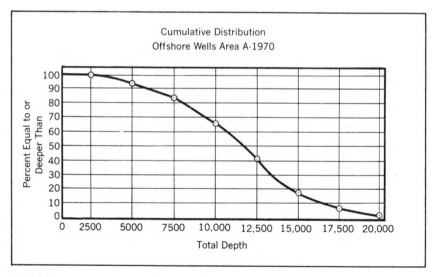

FIG. 1.5

as the two previous figures, but the cumulative percentages begin from the deeper end—to illustrate the "equal to or deeper than" concept. You can read the third point from the bottom as 19% of the wells are in the interval 12,500 to 15,000 ft *or deeper.*

From these figures we can see the first use of a simple mathematical concept to help us understand a mass of data. In the illustration, the mass of data was 365 wells drilled offshore. We examined certain characteristics of these wells as related to depth. Depth was the common classifying point. Grouping the data enabled us to grasp depth relationships quickly. Just imagine trying to gain the same insights from a list of 365 wells and their depths. Your mind would boggle trying to make sense from such a mass of data.

Managing lots of data begins with some form of simplification.

REVIEW

As the human mind grapples with masses of data it searches for meaning. The initial step in this search for meaning is sorting or grouping data by common characteristics. Such groups can be further classified by more specific characteristics. A common mathematical classification is the frequency distribution. In this classification a graph displays either by percentages or actual values the number of times a factor occurs within carefully defined boundaries called class intervals. Cumulative frequency distributions are an important extension of a frequency distribution.

The beginning of understanding of masses of data starts with the proper sorting and classifying.

Recommended Reading

1. Bernstein, Leonard A., "Statistics for the Executive," Hawthorn Books, Inc., New York, 1970.

2 More About Distributions

A brief description of a frequency distribution was in the preceeding chapter. A frequency distribution tells something about a mass of data. It is a step beyond sorting. A frequency distribution is a special arrangement or classification of data. It reduces a mass of data to a few manageable relationships. The 365 wells, for example, were reduced to eight numerical or percentage values.

Within the broad range of possible distributions there are some very specialized distributions. One of these is the Binomial Distribution. Bi—meaning two—and nomial—referring to number, give the clue, then, to a special two-number distribution.

One can think also of the two ingredients in this distribution as two events or two outcomes. For example, you could consider dry or successful for a wildcat well; or you could consider heads or tails in the toss of a coin or coins.

We can learn a lot about distributions, particularly continuous distributions, by creating our own distribution. We can simulate a *continuous* distribution (a smooth curved distribution indicating almost infinite possible outcomes or events) by designing a distribution in which only eleven possibilities can occur. Because only eleven possibilities can occur, our distribution is called a *discrete* distribution. Another discrete distribution would be one in which the variable would be days in a week, months in a year, etc.

A Binomial Distribution

We shall create our distribution with pennies. A single penny tossed can produce only two outcomes—heads or tails. Furthermore with a fair coin, each event is equally likely; i.e. there is a 0.5 probability for a head (50% chance) and a 0.5 probability of tails. Only one event can occur (symbolized as 1.0) so:

The probability (P) of a head (H) plus the probability of tails (T) equals one. In equation form

$$P(H) + P(T) = 0.5 + 0.5 = 1.0$$

Suppose we have ten coins. Let's toss these into the air ten times to see what happens. One toss of the ten coins equals one event. What happened in an actual experiment is shown in Fig. 2.1, Graph *a*. Graph

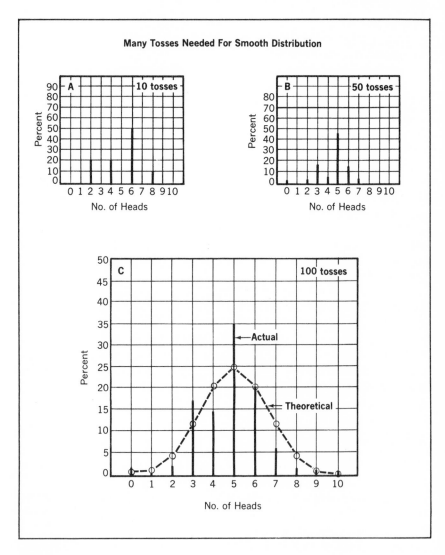

FIG. 2.1

a shows a frequency on the vertical scale (ordinate), and the total possible number of events on the horizontal scale (abscissa). Events are shown as the number of heads which occurred in each toss of ten coins. The number of tails would obviously be 10 minus the number of heads. Note that eleven and only eleven possible outcomes can occur. A toss resulting in no heads is the eleventh possible outcome. These are called discrete probabilities.

Now let's see what happened in ten tosses of ten coins. Amazingly enough, in only 10 tosses *only four* of the eleven possible outcomes occurred. The four events were:

2 heads	twice
4 heads	twice
6 heads	five times
8 heads	once
Total	10 tosses

Please note. The most likely event—5 heads—did not even occur in ten tosses of ten coins.

What happens if you toss the 10 coins fifty times? This result is shown as graph *b* in Fig. 2.1. Now we have *seven events* which occurred. However, we still had, in 50 tosses, four possible events which did not occur. There were none of the 50 events which produced:

1 head
8 heads
9 heads
10 heads

The distribution is a little smoother but still erratic. Note also that the separate session of 50 tosses of ten coins produced no case of 8 heads, whereas the 10 tosses before produced one such outcome.

Graph *c*, Fig. 2.1, shows the results of 100 tosses. Only one of the eleven outcomes did not occur. It was the case of only one head. The distribution is much smoother with the highest frequency being 5 heads (and 5 tails). If each outcome (heads or tails) is equally likely you would expect five heads and five tails to be the most common event.

The dotted line connecting the circles is the theoretical value for ten coins tossed an *infinite* number of times. Thousands of tosses should bring you *very close* to the theoretical values.

The Lessons From Our Manufactured Distribution

What can we learn from this "handmade" series of distributions? The major lesson is in the title of Fig. 2.1. Many tosses are required for a smooth distribution. Let's say it another way. In tossing pennies, a few tosses can produce results markedly different from the average (theoretical) curve generated by many tosses. As we shall see later, a mathematician would state our experience shows that a small *sample* can vary considerably from the total *population.*

Later in this book we are going to relate this idea to wildcat drilling. We will say that in an area with a given success rate, a small number of wildcat wells can produce a number of discoveries which differ markedly from that average success rate.

Note also that even after 100 tosses we don't have actual data which would duplicate a curve of thousands of tosses. The outcome of five heads occurred 35% of the time when ultimately it would only be slightly over 24%. The significance of the number of tosses related to the end result should be apparent from the three illustrations.

Were you surprised that in a large number of tosses five heads would occur only one-fourth of the time? Try this idea on a friend or two. Ask them in a large number of tosses of ten coins what percent of the time you would get five heads! You will be surprised at the range of answers you will get.

Later we have an entire chapter on binomial theorem. There you will learn how to calculate and prove the concepts we just reviewed. It is important to know a little about distributions before we get into the mathematics involved.

Other Values of Probability (*P*)

What if *P* is not 0.5? Suppose you have a condition where the two outcomes are not equally likely? Such a condition is very common in drilling. Development wells usually do better than 0.5 successful, and wildcat wells seldom have a probability of success as high as 0.5.

Fig. 2.2 was drawn to show the theoretical shape of distributions for values of

$$P = 0.2$$
$$P = 0.4$$
$$P = 0.6$$
$$P = 0.8$$

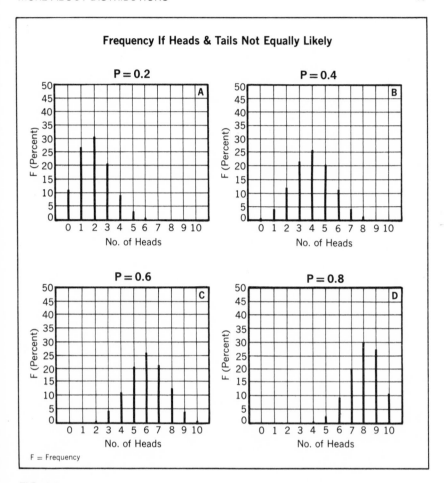

FIG. 2.2

Test tosses were not used here; a large number of tosses is assumed so that the resultant distributions are smooth.

In each case the number of coins (wells) is ten. Graph *a* of Fig. 2.2 shows the theoretical frequency distribution of heads if $P = 0.2$. Here the highest frequency, about 30%, is at two heads as you would expect if $P = 0.2$. Note the similar but reversed shape of Graph *a* to Graph *d*, where $P = 0.8$. If Graph *d* were folded over on top of Graph *a* it would be the same graph! For $P = 0.8$ the most common occurrence is eight heads with a frequency of about 30%.

Note also the reverse similarity of Graphs *b* and *c*. They show the same shape, same frequency distributions, but for different values of heads.

One further observation may escape your eye because of the scales used in the graphs of Fig. 2.2. When $P = 0.5$ a binomial distribution is exactly symmetrical. It is almost symmetrical at $P = 0.4$ and 0.6. This lack of symmetry is barely discernable in Fig. 2.2; but when $P = 0.2$ and 0.8 you can easily distinguish the lack of symmetry. Later a lack of symmetry will be called "skewness."

In Chapter 7 we will learn how to calculate the frequency for a value of P or n (n equals the number of things being measured—in our manufactured example ten coins were used, so $n = 10$). For now we shall just illustrate the principle.

One more fact. It should be obvious that as n increases, the relative frequency of a specific event decreases. For example, if n is 20 ($P = 0.5$), one would not expect ten heads to occur as frequently as did five heads when n was 10. The answer for the probable frequency of ten heads when $n = 20$ and $P = 0.5$ is 17.6%. Logic confirmed! As the number of possible events increases, the chance that any specific event will occur decreases. One might think of an increase in n as an increase in competition among possible occurring events—by virtue of more competitors (events).

The Practical Value of Coin-Tossing

You may say at this point that we have spent a lot of useless time on coin-tossing. Yet there are some very practical applications of a binomial distribution in the real world.

Suppose you have a two outcome condition such as:

1. a defective part
2. death from a disease
3. birth of a boy or a girl
4. hitting or missing a target.

A manufactured part is either good or defective. The outcome from a disease is either life or death. The birth of a child is either a boy or a girl. We have lots of two outcome situations in human life. They will all have binomial distributions.

Suppose a change in the assembly line is introduced to reduce defective parts. The new distribution compared to the old will tell you if the change is working and to what extent. Furthermore, you can tell from a sample rather than a long term run if the new procedure is effective. However, the accuracy of a sample increases with square root of the sample size. Thus to double the accuracy of a sample, it must increase fourfold.[3] We saw the importance of sample size in our coin tossing!

Binomial distributions form the basis for testing the effectiveness of vaccines as well as other types of medical treatment. It also forms the basis for the law of Gambler's Ruin which has certain philosophical applications in exploration. (See Chapter 8).

In truth, however, when n is large (20 or more) and P is not small, the binomial distribution begins to equate to a *normal distribution*. Remember our comment about the lack of symmetry when P was small. We shall save examples of practical uses of distributions until after discussing the *normal distribution*.

The Normal Distribution

The ancient Greeks gave us the first abstract concepts involving numbers. They were the first to divorce numbers from their physical counterparts. They could add $5 + 3$ to get 8 and not worry about what physical things were being added. The Greek concepts of Euclid were involved in absolutely irrefutable concepts. They glorified in proofs which were seemingly inviolate.

Although some pioneering work was started in the 1600's, statistical concepts and their usefulness did not significantly enter the social sciences until the 1800's. The late introduction was partly the result of clinging to the historical tradition, stemming from the Greeks, of deductive reasoning. In the 1800's, many observations about people and their characteristics were found to have a consistent distribution; and some statistical concepts had already been employed in the physical sciences.

Statistical work initiated by John Graunt (1620-74), DeMoiore (1667-1754), and Sir William Petty (1623-75) was revived by a Belgian, L. A. J. Quetelet (1796-1874), in application to the social sciences.[1] Quetelet found, for example, that all mental and physical characteristics of human beings follow a consistent frequency distribution. We now call this distribution the *normal* distribution. Quetelet's measurements of human characteristics included height, size of a limb, weight, head size, intelligence. The physical characteristics of plants and animals also exhibit a normal distribution.

The normal distribution also reproduces the measurements of other physical quantities. Measurements which tend to cluster around a central value exhibit a normal frequency curve or shape. A marksman's variations from a bull's-eye show this tendency. Variations in precise measurements of length, temperature, etc. form such a curve. Because the curve shape shows deviations from a central value it often illustrates the errors from exact measurements. The relationship occurs so frequently that the normal frequency curve is often called the standard error curve—or the normal law of error. It is a most interesting fact; variations of error in measuring physical characteristics do not have some haphazard relationship or chance form—they *always* exhibit the bell-shaped form of a normal distribution.

Fig. 2.3 illustrates a normal distribution. The curve is bell-shaped; and the left half is exactly the opposite of the right half—an important relationship we shall use later.

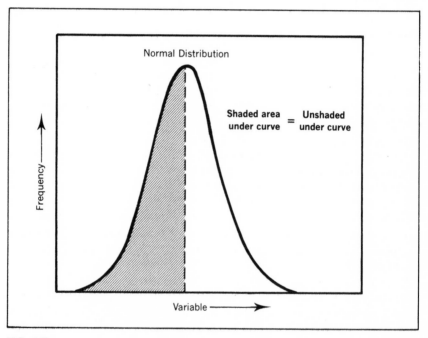

FIG. 2.3

No units are shown in Fig. 2.3 for either scale. The reason is important. The height and width of the curve may vary but as long as the left area (shaded) equals the right area and is the exact opposite in shape, we have a possible normal distribution. So you can have squatty curves, tall ones, or anything in between, yet still possibly have a normal distribution. We say possibly—because not all bell-shaped frequency distributions are normal distributions; but the normal distribution is the most common. That's where it gets its name—normal! This characteristic is important. As we shall soon see, two simple measurements about a normal distribution—the mean and the standard deviation—enable us to describe any normal distribution.

We shall come back to the normal distribution in a later chapter. Next we must learn about some important definitions or characteristics of a distribution.

REVIEW

A particular distribution, the binomial distribution, can be constructed by tossing coins. From this distribution some of the characteristics of a continuous distribution can be inferred and understood.

Many human characteristics when measured give results forming a normal distribution. Under certain conditions the binomial is almost the same as the normal distribution. The normal distribution also illustrates the errors in measurements of physical items, including physical characteristics of human beings.

Recommended Reading

1. Kline, Morris, "Mathematics for Liberal Arts," Addison-Wesley Publishing Co., Reading Mass., 1967, pg. 501.
2. Bernstein, Leonard A., "Statistics for the Executive," Hawthorne Books, Inc., New York, 1970, Chapters 1 & 2.
3. Reichard, Robert S., "The Figure Finaglers," McGraw-Hill, 1974, pg. 109.

3 Defining Characteristics of a Distribution

To this point we have learned a little about distributions. We have glimpsed an application or two which imply usefulness. Before we can proceed further we need to establish and define a few terms which are indispensable to future chapters.

If you already understand the meaning of the terms *mode, median, mean, standard deviation* and *variance,* move to Chapter 4.

Many times when faced with lots of data points we want to learn something about the spread or scatter of the points about some average. The distribution of points about some measure of central tendency is often described by the term "dispersion."

The most important measures of the dispersion of a distribution are the *variance* and its square root, the *standard deviation.*

However, before defining and illustrating these terms, certain others must be reviewed. These other terms define something about a particular "location" in a set of data points. They are the building blocks for understanding variance and standard deviation.

Measures of Central Tendency

The words *sample* and *population* are used by mathematicians to describe two fundamental concepts. The points or values that have been observed from some larger set of data are called a "sample." The larger data set is called the "population." It is the use of a sample to learn about the population (often unknown) that makes statistical concepts such a valuable tool; and there are powerful but simple mathematical techniques which enable us to learn much about an entire population from a relatively small sample.

Three terms we can use to describe a sample are:

1. Median. In any set of data or a "population" there is one point whose value is such that half the remaining points are above and half below. The point whose value exceeds half the data points and is itself exceeded by the other half is called the *median.* Sometimes our data points are observations in an experiment. The median is the middle point of all observations.

2. Mode. In a group of observations if one value occurs more frequently than the other values, it is called the *mode.* The mode is the value of the item measured where the concentration of points is the greatest. We often refer to the measured item as the variable.

3. Arithmetic Mean. The mean is what most people call "the average." It is the sum of all the measured values divided by the number of measurements. It is written in mathematics as follows:

$$\text{Mean} = \frac{\text{sum of values}}{\text{total number of measurements}} \quad \text{or} \quad \bar{X} = \frac{\Sigma(X)}{N} \qquad 3.1$$

If you have twelve watermelons and the sum of their weight is 264 pounds the arithmetic mean is 22 pounds; i.e. when the weight of each is measured and all are totaled, the total divided by the number is the mean.

$$\bar{X} = \frac{\Sigma(X)}{N} = \frac{264}{12} = 22.$$

In statistics the Greek letter Σ (sigma) is always used to mean "sum of." Also $\bar{X}$ is used to represent the arithmetic mean.

There are other types of means (geometric, harmonic, quadratic, root mean square, etc.). One can find references to these in any standard text on statistics. The so-called "weighted average" is in truth a weighted mean. If the word "mean" is used without a qualifying adjective, it is assumed to be the *arithmetic mean.*

The characteristics of frequency distribution just described are shown graphically below:

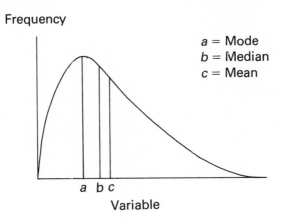

Frequency

a = Mode
b = Median
c = Mean

a b c

Variable

FIG. 3.1

Each characteristic tells a different fact about the data plotted. In general, the mean has many useful properties not possessed by the mode or median.

The distribution shown in Fig. 3.1 is skewed to the right; i.e. it has more values to the right of its highest frequency (mode) than to the left.

Skewness is a measure of the "lopsidedness" of a distribution. This type of curve best illustrates the relationship between mode, median and mean for a non-normal distribution. Under what conditions would the median and mean be on the left side of the mode? In a left skewed distribution the mean and median would be to the left of the mode. The more skewed the distribution the greater will be the mean from the mode. Thus the difference between mean and mode can be used as a measure of dispersion.

A mean can be larger or smaller than a mode. In the right skewed distribution the mean is greater than the mode. In a left skewed distribution the mean will be smaller than the mode.

In a normal distribution the mode, mean and median all have the same value. For this reason it does not illustrate the relationship between the three terms.

Value of These Terms

Of what value are these terms? They are quite useful in risk analysis. To be able to understand clearly and quickly the important relationships about key parameters, expressed as distributions, we need to know, understand, be able to use these terms. For example, much geologic data enter risk simulation programs as triangular distributions. This type of distribution is so important we will have an entire chapter devoted to its use, significance, advantages and limitations.

In a triangular distribution (see Fig. 3.2) you submit three points: a minimum, maximum and a most likely. Frequently the resulting distribution is skewed to the right. In this case the most likely value is a *mode*, not a mean or median. In other words the most likely value will occur more frequently but will not be the average of all values or the middle point. We shall discuss this in much more detail later—and with practical examples. For now you need the assurance that a good understanding of the terms mean, median and mode *is essential.*

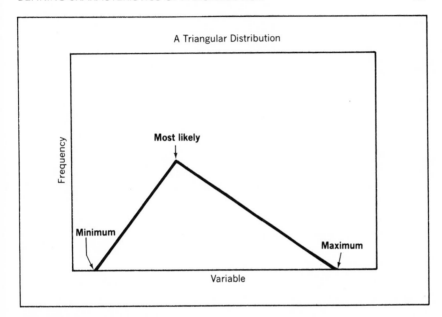

FIG. 3.2

Measures of Variance

Sometimes you want to know more about a set of data than just some characteristic of central tendency (mode, mean, median, etc.). In such instances, other statistical concepts can tell more about the data as a whole. One such concept involves the measurement of variation.

1. Range of Values.   A common measure of variation is the range of values (for example, used to describe sand thicknesses). Ranges express variation with the extremes of the data—the maximum and minimum—as a descriptive device. Knowing only the complete range of values may not give a very satisfactory description of your data. So we have other descriptions of variation.

2. The Mean Deviation.   The mean deviation, a measure of dispersion, is the sum of the deviations of the individual data points from the mean—divided by the number of points. It is, thus, the average deviation from the mean of all data.

Since the sum of the deviations about the mean is zero, ignore signs when summing the deviations. The formula for mean deviation is written:

$$MD = \frac{\Sigma(X_i - \bar{X})}{N} \qquad 3.2$$

where

X_i = specific data point—i.e. each of the data points.

$\bar{X}$ = arithmetic mean

N = total number of data points.

If you calculate a mean deviation it gives you the average difference between all data points and the mean. A large *MD* means wide dispersion—a lack of clearly marked central tendency. A small *MD* shows closely grouped data—not far from the average in value.

The mean deviation illustrates another calculation to give you an insight into a characteristic of a mass of data.

The *MD* may be computed about either the mean or the median. It is most commonly computed about the mean. When the *MD* is computed about the median the value will be smaller because the average deviation from the median is a minimum. Remember the median divides a distribution into two equal areas and is less distorted by unusual values.

3. The Variance. In mathematics the word "variance" is a statistical term, not a descriptive word. The formula for variance (*V*) is:

$$V = \frac{\Sigma(X_i - \bar{X})^2}{N} \qquad 3.3$$

In other words, the variance is

the sum of the squares of the differences between individual values and the mean value

the number of data points

Note the similarity to the mean deviation. The only difference is the *squaring* of each deviation. The squaring causes problems in using the variance. It measures dispersion in squared units and most of us don't think in squared units. For example, the variance for a set of pay thicknesses would have units in feet—squared. Heights of men would have variances in inches—squared. A more convenient measure is our next measure of central tendency.

4. The Standard Deviation. A more convenient unit is the square root of the variance and is called the standard deviation. By taking the square root of variance we will end up with units in feet, inches, acres, etc. The formula for standard deviation is written:

$$\sigma = \sqrt{\frac{\Sigma(X_i - \bar{X})^2}{N}} \qquad 3.4$$

Two symbols are found in text books to denote standard deviation.

a. The lower case sigma "σ" is used for standard deviation when all data points are included—that is, the entire population is represented.

b. If less than the entire population is included, you are dealing with a sample. In this instance the lower case "s" is sometimes used and $n - 1$ (lower case n) instead of N. Both the use of "s" and $(n - 1)$ denote *less than* the total population.

Because all values are squared in the variance the standard deviation is larger than the mean deviation. In a normal distribution the *MD* is about 80% of the standard deviation.

$$MD = 0.798\sigma \quad \text{(for a normal distribution)} \qquad 3.5$$

For moderately skewed distributions this rough relationship still holds approximately true.

The standard deviation is the most popular of all measures of dispersion when dealing with distributions. It gives us a measure of how far data spreads, on both sides of the mean. As the mean is the most important single "describer" of most sets of simple data, so the standard deviation is the most important "describer" of continuous distributions. Remember, both the mean and the standard deviation provide significant characteristics about a set of data. If you know these two facts, you can accurately "describe" the *distribution* of the data. Most present day computer programs dealing with any type of distribution will automatically calculate and print out the standard deviation and the mean.

Like the *MD*, the standard deviation is affected by every single data point. Its values are squared, however. It thus provides greater influence from the maximum and minimum values than the *MD*; even though you eventually take the square root of squared values, the standard deviation produces a value greater than the *MD*.

Calculating a Standard Deviation

The steps in calculating a standard deviation are illustrated using the data in Table 1.1. This table, however, has data that is grouped in depth intervals which complicates matters. We need a simpler example first. For this example we will use the data in Table 3.1. Again, wells and well depths are the basic data set. Please check Table 3.1 carefully. Start with the well depths and see how each column is computed. The symbols $(X_i, \bar{X}, $ etc.) refer to formula 3.4—check it again for reference.

We are now ready to calculate a standard deviation. If you are already familiar with this calculation skip to the next chapter.

TABLE 3.1

CALCULATING THE STANDARD DEVIATION

1	2	3	4	5
	X_i	$(X_i - \bar{X})$		
Well		Dev. From		
No.	Depth	Mean	$(X_i - \bar{X})^2(10^3)$	$(X_i)^2(10^3)$
1	5,000	−4,200	17,640	25,000
2	6,000	−3,200	10,240	36,000
3	7,000	−2,200	4,840	49,000
4	8,000	−1,200	1,440	64,000
5	8,000	−1,200	1,440	64,000
6	9,000	− 200	40	81,000
7	10,000	800	640	100,000
8	12,000	2,800	7,840	144,000
9	13,000	3,800	14,440	169,000
10	14,000	4,800	23,040	196,000
Totals	92,000	24,400*	81,600	928,000

$$\bar{X} = \frac{92,000}{10} = 9,200$$

*Sum with signs ignored.

First let's check the columns in Table 3.1. The first column lists the well number. The second lists the depth for each well—the measured variable. It is the X_i of the formula. The mean (9,200 ft) or $\bar{X}$ is calculated by dividing the sum of the X_i's by the number of wells (N).

The third column shows the deviation of each individual well depth from the mean of 9,200 ft. The fourth column is the square of Column 3. Column 5—$(X_i)^2$—will be used for another calculation later. It is, obviously, the square of Column 2.

Now we fit all of the pieces into Equation 3.4:

$$\sigma = \sqrt{\frac{\Sigma(X_i - \bar{X})^2}{N}}$$

$$\sigma = \sqrt{\frac{81,600,000}{10}} = \sqrt{8,160,000} = 2,857 \text{ ft}$$

The standard deviation of our ten well set of data is 2,857 ft. Another formula to check our answer is:

$$\sigma = \sqrt{\frac{\Sigma(X_i)^2}{N} - \left(\frac{\Sigma X_i}{N}\right)^2} \qquad 3.6$$

This formula uses the sum of Column 5, $(X_i)^2$ from Table 3.1. It also uses the mean—both numbers are squared, but note the different position of N relative to the quantity squared. The latter quantity in the equation $\frac{\Sigma(X_i)}{N}$ is the mean and so one could substitute $(\bar{X})$ for this value. Our new formula becomes:

$$\sigma = \sqrt{\frac{928,000,000}{10} - (9,200)^2} = \sqrt{92,800,000 - 84,640,000}$$

$$\sigma = \sqrt{8,160,000} = 2,857 \text{ ft}$$

How does the standard deviation differ from the mean deviation (MD)? The mean deviation is the sum of column 3, Table 3.1, divided by ten or 2,440 ft. (See formula 3.2.) Remember we said that because the standard deviation used the square root of squared values, it was more sensitive to the extremes. The MD is thus normally less than the standard deviation.

Now we are ready to calculate the standard deviation for a group of wells. Because grouped data are used, the approach must be slightly different. For grouped data we do not have each individual well depth; we have the number of wells in a given depth bracket. With grouped data the steps to calculate a standard deviation are:

1. We find the midpoint of each depth interval. It will be used as the average depth for all wells in the depth interval.
2. A mean is calculated for the entire set of data using the following formula:

$$\bar{X}_g = \frac{\Sigma(n \cdot X_{mp})}{N} \qquad\qquad 3.7$$

Where n is the number of wells in each group, X_{mp} is the midpoint of each depth interval; N is the total number of wells.

3. The mean is subtracted from each midpoint to find the "mean deviation" of each group, $X_{mp} - \bar{X}_g$.
4. Each group or average deviation is squared, $(X_{mp} - \bar{X}_g)^2$. This also conveniently gets rid of negative values.
5. The squared deviation of each group is multiplied by the number (n) of wells in the group, $n(X_{mp} - \bar{X}_g)^2$. This represents the total squared deviation of all the wells in the group.
6. All columns are totalled . . . except the wells and footage which were previously summed to calculate the mean.
7. The standard deviation is then calculated by dividing the total number of wells, N, into the sum of all the group deviations, $n(X_{mp} - \bar{X}_g)^2$, and then taking the square root of the whole shebang. The formula looks like this:

$$\sigma = \sqrt{\frac{\Sigma n(X_{mp} - \bar{X}_g)^2}{N}} \qquad\qquad 3.8$$

We can now construct a table to get a standard deviation from the 365 wells grouped by equal depth intervals.

Inserting our values in equation 3.8 we have the sum of column seven in Table 3.2 divided by the number of wells or:

$$\sigma = \sqrt{\frac{5,597,740,000}{365}} = \sqrt{15,336,273} = 3,916 \text{ ft.}$$

The mean deviation from our 365 wells (using the sum of column eight, Table 3.2,) is:

$$MD_g = \frac{\Sigma n(X_{mp} - \bar{X}_g)}{N} = \frac{1,121,000}{365} = 3,071 \text{ ft.}$$

Again the MD is less than the standard deviation. Most of us tend to think in terms of a mean deviation. To the mathematician, however, the standard deviation, although slightly larger, is more useful particularly for the normal distribution. As we shall see later, he likes it because it is a more powerful tool.

One final point—in grouped data we do not have each individual data point. Groups of data points and their approximately measured

TABLE 3.2

CALCULATION OF A STANDARD DEVIATION-GROUPED DATA

1	2	3	4	5		6	7	8
				Dev. of Midpoint from Mean				
Class Interval	Midpoint (X_{mp})	No. of Wells n	Footage $(X_{mp} \cdot n)$ $\times 10^3$	Actual* $(X_{mp} - \bar{X}_g)$	Squared $(X_{mp} - X_g)^2 (10^3)$	Squared $(X_{mp} - \bar{X}_g)^2$ (10^3)	$n(X_{mp} - \bar{X}_g)^2$ (10^3)	$n(X_{mp} - \bar{X}_g)$ (10^3)
0–2,500	1,250	20	25	−7,775		60,451	1,209,020	155
2,501–5,000	3,750	40	150	−5,275		27,826	1,113,040	211
5,001–7,500	6,250	60	375	−2,775		7,701	462,060	167
7,501–10,000	8,750	100	875	− 275		76	7,600	28
10,001–12,500	11,250	80	900	2,225		4,951	396,080	178
12,501–15,000	13,750	40	550	4,725		22,326	893,040	189
15,001–17,500	16,250	20	325	7,225		52,201	1,044,020	144
17,501–20,000	18,750	5	94	9,725		94,576	472,880	49
Totals		365	3,294	40,000#		270,108	5,597,740	1,121#

$$*\bar{X}_g = \frac{(X_{mp} \cdot n)}{N} = \frac{3,294,000}{365} = 9,025$$

\# Signs ignored.

value are used (in this case depth of the well) with the midpoint of the class interval. Thus the calculated values for *MD* and standard deviation are not absolutely correct. They are, however, accurate enough for almost all uses. In general, the standard deviation and mean deviation will tend to be slightly larger from grouped data than from ungrouped data.

The Mean, Standard Deviation and the Normal Distribution

We have spent much time on the arduous calculations of mean, mean deviation and the standard deviation. We have explained the difference between the mean (or average) deviation and the standard deviation because most of us think in terms of averages. Why have we labored so long? For several reasons:

1. The standard deviation is common output from most computer programs whether summing several distributions or displaying a single distribution. Thus you need to understand the term and how it is calculated.
2. The standard deviation is a vital measure of a normal distribution. With the mean it provides the means to describe and utilize data from any normal distribution. In nature and science so many measurements (or observations) are either normally distributed or nearly so. The mean and standard deviation alone can be used to describe the entire population of a variable or sample with as much accuracy as needed. No wonder the mathematician considers them two very powerful tools.
3. Furthermore many of the measures computed from samples of a population tend to be normally distributed even though the original data are not! The key word here is "tend" as exceptions do occur. The tendency, however, does allow us greater power in extending the usefulness of the mean and standard deviation.

In employing the two terms, mean and standard deviation, the mathematician sets the area under the normal distribution equal to one. Since the sum of any set of probabilities concerning a single event must equal 1.0, portions of the area under the normal curve can be equated to probabilities. We will show how important this concept is when we discuss triangular distributions.

A Standard Normal Distribution

A normal distribution is converted to a *standard* normal distribution by dividing all observations in the *ND* by the standard deviation. This division changes the mean of the *ND* to zero—making it a standard normal distribution. The standard normal distribution has two characteristics:

1) its mean is zero
2) its standard deviation is 1.0.

The ability to convert all normal distributions to a common reference point allows further mathematical manipulation. It provides more answers about our distribution.

The Area Under a Normal Distribution

How do we use the standard normal distribution and its area for probability determinations? First let's look again at a normal distribution.

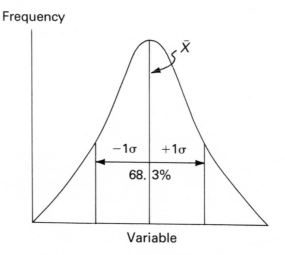

FIG. 3.3

Fig. 3.3 shows a normal distribution. The mean ($\bar{X}$) is the same as the median and mode. One standard deviation is shown on each side of the mean. One of the characteristics of a normal distribution is that the distance equal to one standard deviation measured from the mean includes 34.13% of the total observations. Stated another way, one standard deviation, measured from the mean, will include 34.13% of the area under a normal curve.

Fig. 3.3 shows a standard deviation measured from the mean going both right and left from the mean. Now the area equals 68.3% of the total. Two standard deviations measured off include 95.5% of the area under a normal curve. Three standard deviations would include 99.7% of the area.

Perhaps now you can begin to see the possibilities for our use. Because many natural phenomena and their measurements occur in normal distributions we can use them to help understand problems which have not a single answer but a distribution—i.e., a range of possible answers whose frequencies can be represented as a normal distribution.

How does one use a distribution for probability determination? Suppose we have a problem in which the answer is provided in the form of a *ND*. If two-thirds of the values lie within one standard deviation, plus and minus from the mean, then any value outside (larger than a standard deviation) these limits only has a 33% (one-third) chance of occurring.

Since the area under the normal curve is set equal to 1.0, then portions of that area can be used as portions of 1.0 or as probabilities whose total equals one. That is how the standard deviation, mean and normal distribution produce estimates of probability; and that is why we spent this chapter defining terms relating to distributions.

REVIEW

Estimates of probability often involve distributions. We need, therefore, an understanding of the terms used to describe the characteristics of a distribution. In this chapter the common terms describing a distribution were defined and illustrated. We shall use these terms in subsequent chapters.

Further Reading

1. Wallis, W. A. and Roberts, Harry V., "Statistics, A New Approach," The Free Press, Glencoe, Illinois, 1956, Chapter 11.
2. Mosteller, et al, "Probability and Statistics," Addison-Wesley Publishing Co., Inc., Reading, Mass., 1961.

4 The Lognormal Distribution

One more distribution must be briefly described before we can proceed further. It is the lognormal distribution. In this chapter we will generate a normal distribution and show how the same basic data become a lognormal distribution. We will also begin our discussion of the relevance of this distribution to the search for oil and gas fields.

Generating a Normal Distribution

In Chapter 2 we generated a binomial distribution from tossing pennies. From this manufactured distribution many of the important facts about distributions were observed.

Now we will develop another discrete distribution to illustrate additional aspects about distributions. The example used was suggested by King[1] and deals with the tossing not of coins but dice. Since each die has six possible answers from one toss the number of possibilities exceeds that of a coin. King used four dice in the following way to develop a normal distribution. From each tossing of four dice he summed the faces. Suppose the face values were as follows:

$$3, 4, 2, 6$$

The sum of the four faces equals 15. This value is recorded and the dice tossed again and again and the sum of the faces recorded. The tosses are continued until you have sufficient data to illustrate your example. King used 100 tosses.

What is the range of possibilities here? The lowest sum would be

$$1 + 1 + 1 + 1 = 4$$

The highest sum would be that of the four faces of six, or 24. Our discrete distribution can have values *only* between 4 and 24.

The results of King's 100 throws of four dice are shown in tabular form in Table 4.1 and graphically in Fig. 4.1. It is an uneven distribution as was our 100 tosses of ten pennies; but the resemblance to a normal distribution is indicated.

Note that neither a sum of four nor 24 appeared in 100 tosses of four dice—indicating a rarity of these events. From elementary probability rules we know why this is so. The probability of rolling a 1 from the six possible faces is one in six, or $1/6$.[2] According to the multiplication

TABLE 4.1

VALUES FROM TOSSES OF DICE
THE UNIFORM DISTRIBUTION

Value	Sum	Value	Sum	Value	Sum	Value	Sum	Value	Sum
3 2 6 3	14	3 5 1 3	12	6 3 6 4	19	3 3 1 4	11	3 2 1 3	9
5 5 4 2	16	5 4 6 6	21	1 2 3 4	10	1 4 6 4	15	4 1 4 4	13
4 6 4 4	18	5 2 6 1	14	3 2 6 1	12	6 4 5 3	17	6 2 5 2	15
6 1 3 3	14	5 3 4 6	18	5 4 3 6	18	2 2 2 5	11	1 4 4 2	11
5 1 3 6	15	5 3 2 4	14	6 3 1 6	16	5 3 2 3	13	5 6 4 6	21
5 5 4 6	20	2 5 3 2	12	6 6 4 4	20	5 1 5 5	16	5 5 4 2	18
1 3 6 4	14	4 3 6 3	16	3 5 5 3	16	4 4 4 3	15	2 3 3 1	9
4 6 1 4	15	6 2 6 3	17	3 5 5 5	18	1 4 6 2	13	2 6 2 6	16
2 3 4 5	14	6 2 1 4	13	4 2 5 1	12	1 3 6 2	12	4 6 2 2	14
3 5 5 1	14	2 5 3 2	12	1 1 2 2	6	2 3 5 1	11	6 5 1 1	13
5 5 3 5	18	6 2 6 2	16	5 2 5 3	15	2 6 4 2	14	5 1 3 6	15
6 5 5 2	18	4 2 2 1	9	2 6 3 2	13	1 1 5 1	8	1 1 4 3	9
4 6 6 4	20	1 2 6 3	12	4 6 4 5	19	1 4 2 1	7	2 2 4 1	9
4 1 4 3	12	2 1 1 3	7	3 5 5 4	17	6 2 5 5	18	3 6 1 3	13
2 2 6 2	12	5 2 5 2	14	4 1 1 5	11	5 4 6 5	20	5 5 2 1	13
2 6 5 4	17	2 4 2 2	10	6 6 5 3	20	5 3 4 2	14	2 4 6 5	17
2 3 1 1	7	4 6 4 5	19	6 3 4 4	17	2 1 3 4	10	4 5 4 4	17
3 6 4 3	16	3 1 4 2	10	5 2 6 1	14	6 2 5 1	14	4 6 4 2	16
1 4 5 1	11	2 4 2 5	13	2 6 4 3	15	1 3 3 6	13	1 5 4 2	12
1 3 6 2	12	1 5 4 3	12	2 1 6 2	11	4 1 4 5	14	4 6 5 4	19

After JEK.—With permission from Industrial Press Inc., N.Y., N.Y.

rule the probability of rolling four ones is:

$$1/6 \times 1/6 \times 1/6 \times 1/6 = \frac{1}{1,296}$$

One would expect to roll four ones or four sixes only one time in 1,296 tosses. So it is very normal or natural not to have a value of 4 in 100 tosses. You always have the possibility—it could even happen on the *first* toss; but it is not very probable with only one hundred tosses.

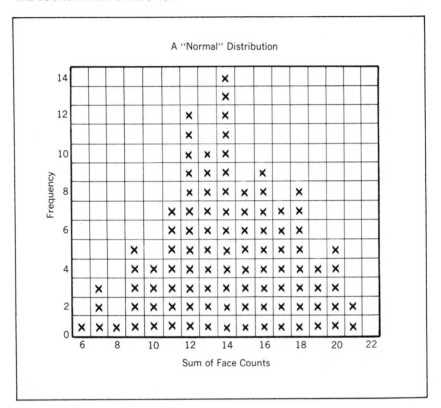

A "Normal" Distribution

FIG. 4.1

The Shape of Our Distribution

Why do the values bunch in the middle ranges of sums? A little deductive thinking supplies our answer. We have shown that the sums of 4 and 24 would be rare—only one chance in 1,296 tosses. All other possibilities then lie between these two values.

Consider the sums of only *two* dice. It is easy to see that the most frequent sum will be seven since it has the greatest number of combinations of two faces, i. e., 1 + 6, 2 + 5, 3 + 4, etc. Seven is the middle value between 2 and 12, the extreme values. Fourteen is the middle value for the combinations between 4 and 24 for four dice; there are more combinations of values which add to this sum than any other.

If seven is the most common sum of the faces of two dice and 14 the most common sum for the faces of four dice, what does this suggest as the most common sum for faces of 8 dice? You can see how a mathematician would take this empirical relationship and develop a formula to fit throws of any number of dice—but that is not our purpose.

Generating a Lognormal Distribution

You can take the same exact data—the raw data—and generate an entirely different distribution. Our raw data are the values on faces of four dice from 100 tosses.

If we take the raw data and *multiply* the four values instead of summing them we get an entirely new distribution—a lognormal distribution. See Table 4.2.

TABLE 4.2

PRODUCTS OF FACE COUNTS FROM TOSSES OF FOUR DICE (RAW DATA GIVEN IN TABLE 4.1)
THE LOGARITHMIC NORMAL DISTRIBUTION

108	45	432	36	18	375	144	150	96	90
200	240	24	96	64	300	16	72	5	12
384	60	36	360	120	576	48	480	8	8
54	360	360	40	32	48	6	300	300	54
90	120	108	75	720	48	100	20	600	50
600	60	576	125	200	240	24	540	120	240
72	216	225	192	18	6	480	288	24	320
96	216	375	48	144	216	24	60	60	192
120	48	40	36	96	20	80	144	54	40
75	60	4	30	30	36	60	24	80	480

After JEK—With permission from Industrial Press Inc., N.Y., N.Y.

We can show the reasons for our difference by the following simple table.

TABLE 4.3

THE FOUR FACES

Sum	Product
$1 + 1 + 1 + 1 = 4$	$1 \times 1 \times 1 \times 1 = 1$
$2 + 2 + 2 + 2 = 8$	$2 \times 2 \times 2 \times 2 = 16$
$3 + 3 + 3 + 3 = 12$	$3 \times 3 \times 3 \times 3 = 81$
$4 + 4 + 4 + 4 = 16$	$4 \times 4 \times 4 \times 4 = 256$
$5 + 5 + 5 + 5 = 20$	$5 \times 5 \times 5 \times 5 = 625$
$6 + 6 + 6 + 6 = 24$	$6 \times 6 \times 6 \times 6 = 1,296$

The range of values for a sum was from 4 to 24; but for products the range is from 1 to 1,296—a much broader range of values.

The new range of values is so broad that we have to group our data to comprehend the results better. As we stated in Chapter One, grouping data properly allows us to condense much information into small enough units for easier comprehension.

On Fig. 4.2 we have done just that. We have grouped the products into ranges of 100. The marked difference in the two relationships is

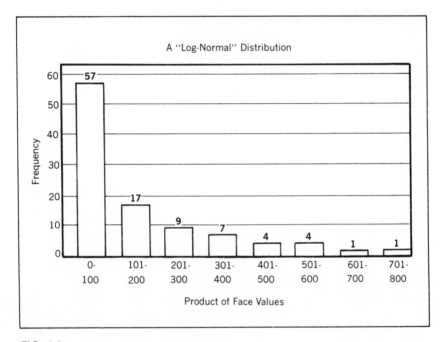

FIG. 4.2

apparent. The product values bunch at the low end of the distribution. Why? Examine Table 4.3 again. It is just as common to get face values of 1, 2 and 3 as it is to get values of 4, 5 and 6. Products with one, two and three should have low values mostly below 100; and that is exactly what Fig. 4.2 says. In fact, 57 of the 100 tosses resulted in products of 100 or below. Note the sharp decrease in the number of products of higher values.

No product above 800 was rolled, even though the top range is 1,296. However, as shown for the normal distribution the chance of getting four faces with the value of six is rare.

A Natural Lognormal Distribution

Now let's look at some natural relationships that have a lognormal distribution. Many relationships involving money exhibit a lognormal distribution. Family income is one. Although the number of millionaires increases with time, percentagewise they are a small part of the number of income units.

Church pledges exhibit a lognormal distribution. On Table 4.4 the pledges from a suburban church are listed. The 296 pledges received were grouped by 100 dollar class interval—the interval representing the annual pledge. Column 2 lists the number of pledges in each interval. Column 3 is the percent each interval is of total pledges; and Columns 4 and 5 show the cumulative percentage, Column 4 beginning with the smallest size building to 100%, and Column 5 beginning with the largest

TABLE 4.4

A DISTRIBUTION OF CHURCH PLEDGES

	1	2	3	4	5	6	7
				Percent		Midpoint of	
	Class		Of	Cumulative*		Class Interval	
Row	Interval	Actual	Total	Small	Large	Midpoint	Log_{10}
1	0–99	18	6.1	6.1	100.0	50	1.70
2	100–199	74	25.0	31.1	93.9	150	2.18
3	200–299	41	13.9	45.0	68.9	250	2.40
4	300–399	57	19.2	64.2	55.0	350	2.55
5	400–499	26	8.8	73.0	35.8	450	2.65
6	500–599	22	7.4	80.4	27.0	550	2.74
7	600–699	11	3.8	84.2	19.6	650	2.81
8	700–799	14	4.8	89.0	15.8	750	2.87
9	800–899	6	2.0	91.0	11.0	850	2.93
10	900–999	7	2.4	93.4	9.0	950	2.98
11	1,000–1,099	6	2.0	95.4	6.6	1,050	3.02
12	1,100–1,199	1	0.3	95.7	4.6	1,150	3.06
13	1,200–1,299	3	1.0	96.7	4.3	1,250	3.10
14	1,300–1,399	4	1.4	98.1	3.3	1,350	3.13
15	1,400–1,499	—	—	98.1	1.9	1,450	3.16
16	1,500–1,599	3	1.0	99.1	1.9	1,550	3.19
17	1,600–1,699	—	—	99.1	0.9	1,650	3.22
18	1,700–1,799	—	—	99.1	0.9	1,750	3.24
19	1,800–1,899	—	—	99.1	0.9	1,850	3.27
20	1,900–1,999	1	0.3	99.4	0.9	1,950	3.29
21	2,000–2,099	1	0.3	99.7	0.6	2,050	3.31
22	2,100–2,199	—	—	99.7	0.3	2,150	3.33
23	2,200–2,299	1	0.3	100.0	0.3	2,250	3.35
	Total	296					

*i.e. beginning with the smallest or largest class interval to build to 100 percent.

size. Columns 6 and 7 deal with plotting relationships used later. They contain the midpoints of the class interval and the logarithm to the base 10 of the midpoint value.

A quick glance at Table 4.4 shows most pledges bunched at the lower end—just exactly like the products of the four faces were in Fig. 4.2! We have an approximate lognormal relationship!

Looking at church pledges gives you a simplified definition of a lognormal distribution—

"Lots of little values and not many big ones!"

Column 4 is particularly revealing. Note these facts:

1. 45% of the individual annual pledges are less than $299.00.
2. About 65% are less than $399.00.
3. 80%—less than $599.00.
4. Only 5% are greater than $1,100.00 per year.

A linear plot of the pledges is shown in Fig. 4.3. Note the similarity in shape to Fig. 4.2. Now check Fig. 4.4. The horizontal scale shows

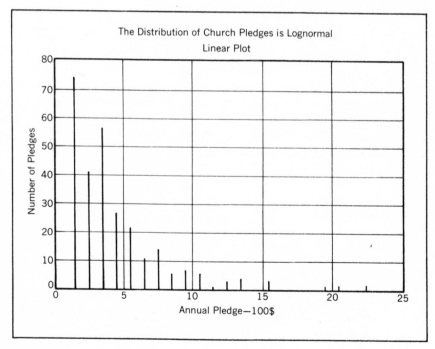

FIG. 4.3

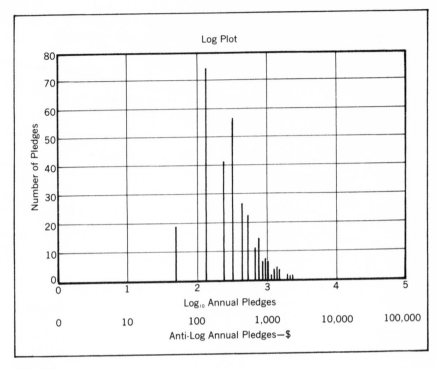

FIG. 4.4

the *logarithm* of annual pledges, not the pledge value itself. Look at the change in shape. The pledges now have a shape about like a normal distribution—Fig. 4.1.

Herein lies the definition of a lognormal distribution. If the logarithm of a variable vs. frequency plots as a normal distribution (assumes the shape of the normal curve) you have a lognormal distribution.

Log Probability Paper

One of the easy ways to check for a lognormal distribution is through a special kind of graph paper. It is called log probability paper. Data which plots on this paper as a straight line can be considered lognormal. In the most simple form you need individual data points because you must have the size associated with a particular (individual) percent of the total sample. Data on Table 4.4 are grouped and do not lend themselves easily to such a plot.

TABLE 4.5

AN APPROXIMATE LOGNORMAL DISTRIBUTION

Bank Rank	Deposits—M$* As of 6-30-74	Fractile %
1	448	3.6
2	301	7.1
3	158	10.7
4	122	14.3
5	95	17.9
6	92	21.4
7	84	25.0
8	83	28.6
9	80	32.1
10	71	35.7
11	61	39.3
12	52	42.8
13	46	46.4
14	35	50.0
15	33	53.6
16	30	57.1
17	24	60.1
18	23	64.2
19	18	67.8
20	10	71.4
21	10	75.0
22	7	78.5
23	6	82.1
24	5	85.7
25	3	89.3
26	2	92.8
27	1	96.4
Total	1,900	

*Savings & Loan institutions, Harris County, Texas.

Fortunately for a simple illustration alternate data are available. On Table 4.5 are listed the deposits in savings and loan institutions in Harris County, Texas. They are ranked by size and listed in decreasing order. The largest institution has almost $450 million in deposits. The cumulative fractile percentages are shown in Column 3. Fractile percentages are calculated by using the total sample + 1 as the divisor.[2] In the data set a single fractile is $\frac{1}{27 + 1}$ or .0357 (3.6%); i.e. each data point is about 3.6% of the total sample.

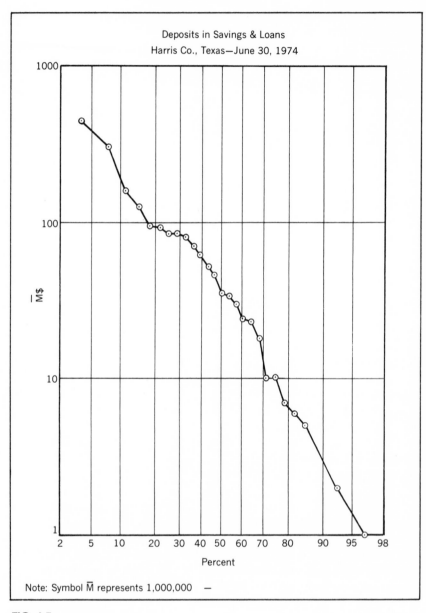

FIG. 4.5

Data from Table 4.5 are plotted in Fig. 4.5. The vertical scale for the variable is logarithmic; the horizontal scale for the percentages is a special one which stretches out the data points. We do not have an absolutely straight line relationship for the data; therefore, the distribution is only approximately lognormal. Nevertheless a general lognormal trend is visable.

A Short Cut

Suppose you have a large data sample—100 or more items. You may not want to plot all 100 items to check for lognormality. A short cut exists.

There were 136 banks in Harris County, Texas, on June 30, 1974. Their deposits were listed, ranked with fractiles calculated as in Table 4.5 for savings and loan institutions. The result is plotted on Fig. 4.6. You do not see all 136 points, however. Only twelve are shown. Those shown represent the data from every 13th bank in decending order with one point added at 5% for further curve definition.

When faced with lots of data, plot every 10th, 20th etc. point. The number to plot depends on the population size—number of data points.

Note our data again do not form an absolutely straight line. There is a tail upward at the top, away from an otherwise good trend. So again we have only an approximate lognormal relationship.

Two banks had deposits above one billion dollars. They do not appear on the graph paper—falling just outside the 2% line. This omission illustrates one fact about some types of log probability paper. Much of it stops at 2% on the high side and 98% on the low side. For a large sample the extreme points are thus not shown. But frankly, the extremes are not what you are usually interested in when using a distribution. They are the atypical, not the typical.

Lognormal Distributions and Oil & Gas Fields

An entire chapter has been devoted to the lognormal distribution—for sound reasons. It is the most important distribution in the search for hydrocarbons. It has been well documented that reserve sizes of oil and gas fields in a given play or basin form a lognormal distribution.[3,4] There are few giant fields and many small ones. We need, therefore, to understand this particular size arrangement in nature so we can use it effectively.

Why should oil and gas fields show a lognormal distribution? Geologi-

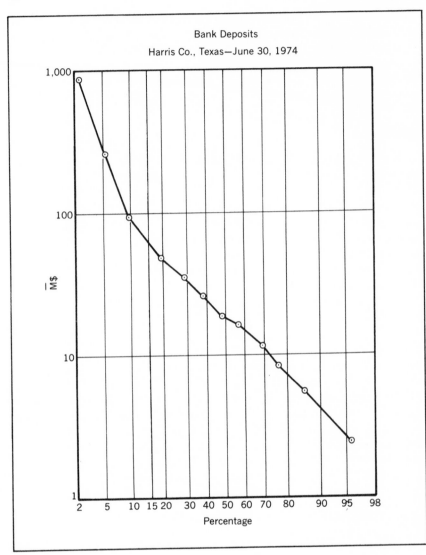

FIG. 4.6

cally speaking they do because of the rarity of the special conditions in nature required to produce large fields.

If you need a mathematical answer, the example of products (from the 4 die faces) forming a lognormal distribution should suffice. Estimates of reserve size will almost always involve multiplication. The example used in Exploration Economics[2] multiplied:

net pay × recovery per acre × area.

Logically then estimates of reserve sizes should line up in a lognormal distribution because they arise from multiplication rather than adding.

REVIEW

The same data which build a normal distribution can be used to construct a lognormal distribution. Adding random data produces a normal distribution; multiplying random data produces a lognormal distribution. Much statistical data involving money and its derivatives (incomes, deposits, etc.) show lognormality. Log probability paper offers a quick way to check a data set for lognormality. The lognormal distribution has special significance to the explorer. It illustrates the way the sizes of oil and gas fields are distributed. Our simplified description of a lognormal distribution is:

"Lots of small values and not many big ones."

BIBLIOGRAPHY

1. King, James E., "Probability Charts for Decision Making," Industrial Press, Inc., New York, 1971, pp. 66-68.
2. Megill, Robert E., "An Introduction to Exploration Economics," The Petroleum Publishing Co., Tulsa, Oklahoma, 1971, pp. 112-113.
3. Kaufman, Gordon M., "Statistical Decision and Related Techniques in Oil and Gas Exploration," Prentice-Hall, Inc., Englewood Cliffs, N.J., 1963.
4. McGrossman, R. C., "An Analysis of Size Frequency Distributions of Oil and Gas Reserves of W. Canada," Canadian Journal of Earth Sciences, Vol. 6, 1969, pp. 201-211.

5 The Shape of Distributions

We have devoted four chapters to enhancing an understanding of distributions. We have checked a few of the mathematical properties of distributions and tried to convey meanings about some of their characteristics.

We need the ability to make a quick mental leap from a basic frequency histogram—a distribution—and its resultant shape in the cumulative form. Why? The cumulative frequency distribution is often used in risk analysis. Therefore, we need to be able to think both forward and backward—forward from a histogram to a cumulative frequency and backward from a cumulative curve to the input histogram. Seeing one we need to be able to envision the other.

This chapter will help develop the transition to easy recognition of distribution shapes.

Two Methods of Display

To begin with, let's review what we saw in Figs. 1.4 and 1.5, where we plotted a distribution for offshore wells. There we saw two different means of displaying a cumulative frequency distribution—each with the same vertical scale.

Fig. 5.1 repeats the general relationship for these two presentations. The earlier curves, referring to well depth, used the terms

—equal to or shallower than
—equal to or deeper than.

In other usages, with parameters other than well depth, we will see the terms "less than" and "greater than" with shapes as shown in Fig. 5.1. The data involved are the same and the same information can be gleaned from each graph. For a given parameter size one curve "says" 25% are equal to or less than that value. For the same size the "greater than" curve says the same thing in different words. It says 75% of the parameter sizes are equal to or greater than that value.

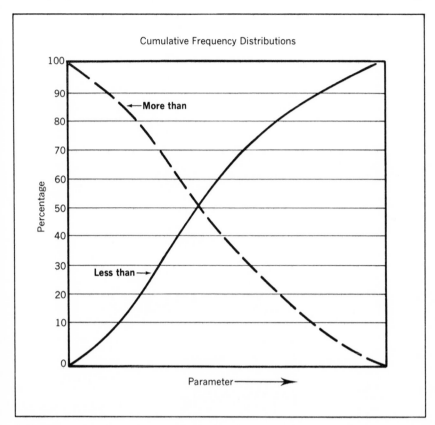

FIG. 5.1

Distribution Shapes

On Fig. 5.2 eight different frequency histograms are shown—A through H. Each also has to its right the resultant cumulative frequency distribution. All numerical values shown are percentages. A brief review of these shapes will vastly enhance our ability to recognize frequency distribution shapes when viewing only the cumulative curve. The cumulative curves are drawn in the "greater than" shape.

Distribution A

We begin with A. It is a rectangular distribution. What is a rectangular distribution? A rectangular distribution is one in which each parameter

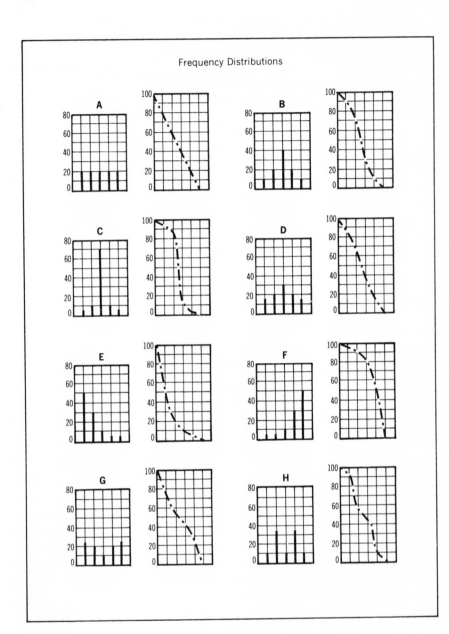

FIG. 5.2

value (of which there are five) has an equal frequency of occurrence. A typical rectangular distribution would be the frequency of rolling 1, 2, 3, 4, 5 or 6 with one die. Each face has an equal chance of occurring.

Now note the cumulative curve to the right. It is a *straight line*. *Thus any straight line represents a rectangular distribution;* and a rectangular distribution occurs when each outcome or value has an equal chance of happening. Straight lines mean equal frequencies.

Distribution B

Distribution B has a shape similar to a normal or bell-shaped curve. It produces a curved cumulative frequency distribution. Note the steep slope in the center. It corresponds to the highest frequency of the B distribution. Here our second point emerges—*steep slope means high frequency.*

Distribution C

Our second point appears more pronounced in the C distribution. Note the very sharp, steep slope—again related to the highest frequency. Steep slope relates to your "most likely" occurrence. Seventy percent of the parameter values in C fall within the middle value. You can see this in the C distribution. Can you see it also in the cumulative curve? If so, how? You see it in the cumulative frequency curve by noting that the steep slope occurs between 85% and 15%. Thus 85 minus 15 shows 70% of the values occur in a very narrow range of frequencies—represented in the C distribution by the tall bar.

Distribution D

D is of special interest. Check its frequency histogram. The middle value is largest, but not so pronounced as in B and certainly not like C. In fact, D is between A and B in shape. Its cumulative curve demonstrates this fact well. It is not a straight line but less curved than B's cumulative. With no major parameter differences D has no detectable sharp slope on the cumulative curve. *No sharp slope means no sharply differing frequencies.* Do you begin to see the shapes and their relationships? Can you see D's cumulative curve and say "that's almost a rectangular distribution!" If so, you are getting the message on shape recognition.

Distribution E

As we shall see later, E is somewhat akin to a lognormal distribution. It is skewed to the right. If it were more right skewed it would be approximately lognormal. It does show mostly small values and few big ones. The cumulative curve shows steep slope on the left—flattening to the right.

When N (the number of items) is large and P (the probability of occurrence) is small, the Poisson distribution has this shape; the Poisson distribution results in situations where an event can occur more than once, for example, automobile accident experiences over a period of time. The binomial distribution has this shape when P is small.

Cumulative distributions for oil and gas fields should always have their steep slope to the left. Nature made more small fields than giant ones. Remember the shape of E.

Distribution F

F is the opposite of E. As such it gives an "upside down" cumulative curve relative to E. One would never expect to find F in a size distribution of oil and gas fields.

Distribution G

G is a peculiar distribution. It represents a situation where the highest values are at the extreme ends of the distribution. As you see, the steep slopes of the cumulative curve are there also.

Distribution H

H shows a bimodal distribution—two peaks of about the same frequency. It is not uncommon in natural statistical gatherings. Again steep slopes go with high frequencies. Can you read from H's cumulative curve the percentages associated with the highest frequencies? Use the same method employed in the review of C.

Approximate Lognormal Distributions

Distributions I and J

Distributions I and J on Fig. 5.3 show shapes akin to E but more toward lognormality. Note the long tail on J. You would think it is surely lognormal. How can we check?

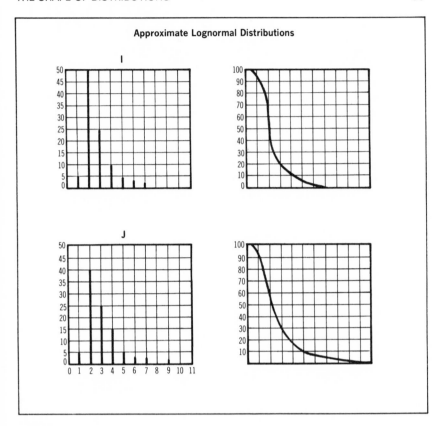

FIG. 5.3

The easiest way is to plot the data on log probability paper. Remember we said it is a special kind of paper deliberately set up so that lognormal distributions plotted on it will show as straight lines.

The following steps enable a plot on lognormal probability paper:[1]

1. Arrange the parameter values by size starting with the largest value first.
2. Calculate the cumulative percentages for each observed point. In Table 5.1 (See curve J, Fig. 5.3) there are nine parameter values. Each is thus 11.1% (1/9 of 100%) of the total sample. These percentages are known as *percentiles*.
3. Find the midpoint of each percentile for each value on the cumulative distribution. This is done by taking half the first

percentile $\left(\dfrac{11.1\%}{2}\text{ or }5.5\%\right)$ for the first point, and adding the full percentile (11.1) to the first point to get the second (11.1 + 5.5 = 16.6). The third point is obtained by adding 11.1 to 16.6, etc. Fractiles, rather than midpoints, were illustrated in the previous chapter. Both techniques are used and essentially the same plot results for 10 or more points. Midpoints are probably easiest to use and understand.

4. Plot the midpoints versus the parameter values.

The following Table, 5.1, shows the raw data for curve J.

TABLE 5.1

RAW DATA—CURVE J

(1)	(2)	(3)
Parameter Size	Cum. Percent	
	Actual	Midpoint
11	11	5.5
9	22	16.5
7	33	27.5
6	44	38.5
5	55	49.5
4	66	60.5
3	77	71.5
2	88	82.5
1	99	93.5

It takes Column 1 and Column 3 to plot a line on lognormal probability paper. The size (11) is plotted versus midpoint of the cumulative percentage (5.5). These two columns are plotted on Fig. 5.4 as the lower line J.

J does not form a straight line. Therefore, it is not lognormal. The largest values for J would have to be larger for a straight line and a lognormal distribution.

Note the straight line K. It represents a truly lognormal distribution. Its raw data are:

The straight line, K, on Fig. 5.4 was plotted from Columns 1 and 3 as in the case of J. What would K look like on rectangular coordinate paper? The answer is on Fig. 5.5. Note the long tail—strong skewness to the right.

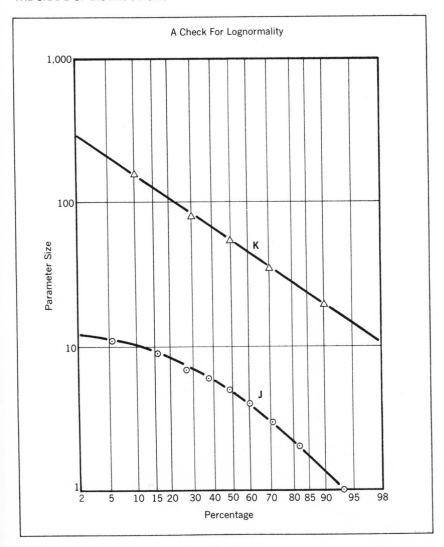

FIG. 5.4

REVIEW

Risk analysis deals in probabilities and probabilities often are expressed in the form of either frequency histograms or as cumulative frequency distributions. It is important, therefore, to be able to recognize the meaning of a given cumulative curve. Each distribution has its own particular resultant

TABLE 5.2

RAW DATA—CURVE K

(1) Parameter Size	(2) Cum. Percent Actual	(3) Midpoint
160	20	10
80	40	30
55	60	50
35	80	70
20	100	90

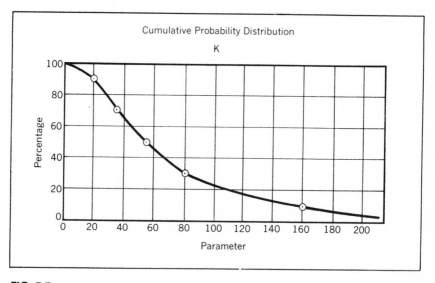

FIG. 5.5

cumulative shape—from the straight line of a rectangular distribution to the sloping curve of the lognormal distribution.

Steep slope always means high frequency. Thus, for curves of cumulative probabilities, steep slope expresses a higher degree of certainty or occurrence. Flat slopes express uncertainty. These simple concepts will have value as we proceed further.

BIBLIOGRAPHY

1. Megill, R. E., "An Introduction to Exploration Economics," The Petroleum Publishing Co., Tulsa, Okla., 1971, pg. 113.

6 Permutations and Combinations

We are well on our way to the completion of Part I. Yet there are still important mathematical concepts needed for an easy understanding of principles in risk analysis. One such concept involves permutations and combinations. These two words are building blocks to an understanding of binomial theorem which in turn allows an insight into the law of "Gambler's Ruin." We shall develop an understanding of permutations first.

Permutations

Permutation is a fancy word for the arrangement of things—but it is a special kind of arrangement in that *order* counts. What do we mean when we say order counts? Each different order represents a different permutation.

For example, in how many ways can two books be arranged *in order* on a shelf? If we call the books "a" and "b", the answer is:

ab
ba

or two ways.

We have the same two books, but they are arranged in two different orders and, therefore, count as separate permutations. Order counts!

Suppose we have three books, a, b and c—how many ways can they be arranged in order on a shelf? Our answer is:

abc bca
acb cab
bac cba

or six ways.

Again we can see the significance of *order*. Note that abc and acb count as two separate arrangements. Changing the order makes a different arrangement.

You can use a tree diagram to illustrate the arrangement of things. Such a diagram for the prior example would be:

1st Space	2nd Space	3rd Space	Possible Arrangements
a	b	c	abc
	c	b	acb
b	a	c	bac
	c	a	bca
c	a	b	cab
	b	a	cba

The zero point is shown so you have a common point to count from. Starting at this common point, you can move to the right and see all possible arrangements of three items when order counts.

Now suppose we have four items? In how many ways can four books be arranged *IN ORDER* on a shelf? If we call the books a, b, c, and d, we have the following possible arrangements:

abcd	bacd	cabd	dabc
abdc	badc	cadb	dacb
acbd	bcad	cbad	dbac
acdb	bcda	cbda	dbca
adbc	bdac	cdab	dcab
adcb	bdca	cdba	dcba

There are 24 ways four items can be arranged if order counts. You can see that drawing tree diagrams or listing all possible combinations gets very tedious, if not impossible, for a large number of items.

Fortunately, mathematicians have developed a shortcut for this calculation. Mosteller[1] et al. use squares to indicate the positions on the shelf—like this, for four items:

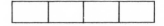

To fill in the spaces you consider some of the possibilities and limitations. In the first space we can put any one of the four books.

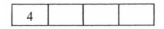

Once we do this we limit the possibilities in the second space. Any one book placed in the first space leaves only three possibilities for the second space.

4	3		

Following this logic, we see that for the third space we only have two books left—two possible choices.

4	3	2	

For the last space, after the three other choices are made, there is only one book left and we fill in the last space with 1.

4	3	2	1

Remember when we listed all possible arrangements for four items it came out 24? We can get the same answer by multiplying the four numbers in the boxes times each other.

$$4 \times 3 \times 2 \times 1 = 24$$

This number corresponds to the 24 branches that would appear on a tree diagram illustrating the example of four items.

Using this formula we can check the ways in which five things can be arranged in order:

$$5 \times 4 \times 3 \times 2 \times 1 = 120$$

Trying to show how five things could be arranged in order by a tree diagram would be very time-consuming. We can be grateful to the mathematicians for giving us this shortcut.

A Formula for Permutations

In developing a formula for permutations, mathematicians use a special symbol. It is called the factorial symbol. The factorial symbol denotes the product in declining order of whole numbers, *beginning with* the number shown. The symbol is indicated by the exclamation point. For example:

$$1! = 1$$

$$2! = 2 \times 1! = 2$$

$$3! = 3 \times 2 \times 1 = 3 \times 2! = 6$$

$$4! = 4 \times 3 \times 2 \times 1 = 4 \times 3! = 24$$

$$5! = 5 \times 4 \times 3 \times 2 \times 1 = 5 \times 4! = 120$$
$$6! = 6 \times 5 \times 4 \times 3 \times 2 \times 1 = 6 \times 5! = 720$$

$$\cdots$$

$$7! = 5040$$
$$8! = 40{,}320$$
$$9! = 362{,}880$$
$$10! = 3{,}628{,}800$$

You can see that a factorial number can express a large number;

$$10! = 3.6 \times 10^6$$
$$26! = 4 \times 10^{26}$$

and that the numerical value of factorials increases at a tremendous rate.

Perhaps you have already guessed why we introduced the factorial symbol. It is the formula for simple permutations. We have seen that the permutations of four different things, taken all together (order counts) is 4!. The general equation has the form of:

$$_4P_4 = 4!$$

The number of permutations (arrangements) of "n" things taken all together is $n!$. This expression is usually written:

$$_nP_n = n!$$

It is read as the permutations of n things taken n at a time (all together) equal $n!$.

But suppose we don't use all n items; i.e., we don't take them all together. As an example, imagine that we want to know the number of arrangements which can be taken from 6 things taken three at a time. Our equation is written:

$$_6P_3 = \frac{6!}{(6-3)!} = \frac{6!}{3!} = \frac{720}{6} = 120$$

Another way to evaluate $_6P_3$ is to begin with 6 and proceed for three numbers:

$$_6P_3 = 6 \times 5 \times 4 = 120$$

This method is even more simple for this illustration than the use of factorial numbers. Why does it work? It works because 6! is really $6 \times 5 \times 4 \times 3!$, and therefore,

$$_6P_3 = \frac{6 \times 5 \times 4 \times 3!}{3!}$$

and the 3! in the numerator and denominator of the fraction cancel each other and

$$_6P_3 = 6 \times 5 \times 4 = 120$$

The equation which expresses this relationship for all values of n and x is:

$$_nP_x = \frac{n!}{(n-x)!}$$

It is read—the number of permutations of n things taken x at a time equals:

$$\frac{n!}{(n-x)!}$$

Another example would be the number of arrangements of 5 things taken two at a time:

$$_5P_2 = \frac{5!}{(5-2)!} = \frac{5!}{3!} = \frac{5 \times 4 \times 3!}{3!} = 5 \times 4 = 20$$

Again, to use the short form of $_5P_2$ you begin with 5 and proceed 2 factors:

$$_5P_2 = 5 \times 4 = 20$$

Just to show that our equations still represent the world of reality, let's construct an example of $_5P_2$ (i.e. the permutations of 5 things taken 2 at a time). We can again assume books for our physical items—and indicate them by the symbols a, b, c, d and e. Remember however that our number of arrangements can only involve two books at a time even though there are five books available. Under this constraint our arrangements are:

ab	bc	cd	de	ca
ac	bd	ce	da	eb
ad	be	ca	db	cc
ae	ba	cb	dc	ed

Thus we have shown by our example that five things taken only two at a time ($_5P_2$) can make exactly 20 possible arrangements; we have confirmed that

$$_5P_2 = 20.$$

Repetitions

The game changes slightly if you can have repetitions. In a prior example of four books, arranged in order, we had 24 possible arrangements. Suppose you had four copies of each of the four books and could make repetitions. Then in each space you could always put one of each book; and one possible arrangement would be all four of the same book—aaaa. The possible arrangements with repetitions would be:

$$4 \times 4 \times 4 \times 4 = 256$$

Repetitions Expand the Possibilities. One can also use permutations with repetitions to consider different things arranged together. A common example used in textbooks is auto license plates. Most states use two things on their plates—numbers and letters. When cars were not so plentiful, numbers were sufficient. If you assume you don't want a zero in the first space, a five-number license could show the maximum different arrangements for the 10 possible numbers (0 through 9) as follows:

$$9 \times 10 \times 10 \times 10 \times 10 = 90,000$$

With letters, however, you can expand the number of different arrangements, since 26 different letters can go into the first space as follows:

$$26 \times 9 \times 10 \times 10 \times 10 = 234,000$$

Just the addition of a letter in the first position, in a five-position license, changes the possible arrangements from 90,000 to 234,000. Two letters and three numbers expand the possible arrangements even further:

$$26 \times 26 \times 9 \times 10 \times 10 = 608,400$$

Three letters and two numbers produce:

$$26 \times 26 \times 26 \times 9 \times 10 = 1,581,840$$

We can illustrate the effect of repetitions a different way when numbers and letters are not mixed. Consider:

$$_4P_2 = 4 \times 3 = 12$$

We illustrate the formula with a, b, c, d as follows:

ab	ba	ca	da)
ac	bc	cb	db) = 12
ad	bd	cd	dc)

But with repetitions we can add aa, bb, cc, and dd or four more possibilities.

$$12 + 4 = 16$$

or 4 items taken 2 at a time $(_4P_2)$

but with repetitions $= 4 \times 4 = 16$. Then n items taken x at a time *with repetitions* equals n^x.

Combinations

Our next building block involves combinations. A combination is an arrangement (or sub-set) in which *order does not count*. It thus differs from a permutation where order does count.

In a previous example, we saw that

$$_5P_2 = 20$$

Now suppose we say that order does *not* count. How many combinations do we have? Assume that we have five things, a, b, c, d, and e. Then possible combinations are:

ab	bc	cd	de
ac	bd	ce	
ad	be		
ae			

There are ten combinations. Note, we cannot have ba because it is the same as ab *if* order does not count.

Let's take another previous example. We saw that:

$$_6P_3 = 120 \text{ permutations}$$

How many combinations are possible from six things taken three at a time? Remember order does not count! Consider our things to be a, b, c, d, e, and f. Our combinations are:

abc	bcd	cde	def
abd	bce	cdf	
abe	bcf	cef	
abf	bde		
acd	bdf		
ace	bef		
acf			
ade			
adf			
aef			

There are twenty combinations. Again we cannot count efa as a separate combination because it is the same as aef, since order doesn't count.

It seems obvious that under *most* circumstances there will always be fewer combinations (order is ignored) than permutations (order is everything!). We might also suspect that there is some relationship between a combination and a permutation—and there is. Our general formula for permutations was:

$$_nP_x = \frac{n!}{(n-x)!}$$

The general equation for combinations is

$$_nC_x = \frac{n!}{x!(n-x)!}$$

The difference is just an $x!$ in the denominator and we can prove it by calculating the two samples we counted.

Combinations of 5 things, taken 2 at a time, are:

$$_5C_2 = \frac{5!}{2!(5-2)!} = \frac{5!}{2!\,3!} = \frac{5 \times 4 \times 3!}{2!\,3!} = \frac{5 \times 4}{2!} = \frac{5 \times 4}{2 \times 1} = 10$$

Combinations of 6 things taken 3 at a time are:

$$_6C_3 = \frac{6!}{3!(6-3)!} = \frac{6!}{3!\,3!} = \frac{6 \times 5 \times 4}{3!} = \frac{120}{6} = 20$$

Now we have a new formula for determining combinations. To understand its relationship to permutations consider what happens to the equation for combinations

$$_nC_x = \frac{n!}{x!\,(n-x)!}$$

if we multiply the equation times $x!$ It then becomes $_nC_x(x!) = \dfrac{n!}{(n-x)!}$;
the right side is now the formula for permutations. Thus the number of combinations (C) times the factorial $(!)$ of the number taken at a time (x) equals the number of permutations (P).

Some authors use a different notation for combinations. Instead of writing $_nC_x$ they simply write $\binom{n}{x}$. So if you see $\binom{5}{2}$ or C_2^5 or $C\binom{5}{2}$ these are not fractions but expressions meaning the same as $_5C_2$. Springer et al.[2] use an even different notation. They use a lower case c followed by (n, x); e.g.

$$c\,(n, x) = {}_nC_x = C_2^5$$

Mathematicians need a bureau of standardization to encourage all to use the same symbols!

There are many practical uses for combinations and permutations. License plates were mentioned earlier; but one of the most ancient and *still* useful application of permutations is in the construction of locks. In fact Jean Borrel wrote about the significance of permutations and locks as early as 1559!

Suppose you have a cylinder for a lock with five pins. How many different key combinations are possible? You have to make an assumption about the number of different positions an individual pin in the lock's cylinder can assume. The usual number is five. Using your knowledge of permutations determine the number of possible combinations for a 5 pin lock. Calculate the number of different permutations for six pins in a cylinder and you can see why more pins mean more security. One additional pin increases the number of possible combinations several fold!

REVIEW

Permutations and combinations represent key building blocks as we move toward concepts in risk analysis.

In later chapters an understanding of these concepts will "open the door" to uses of binomial theorem. They also provide an understanding with binomial theorem of the law of Gambler's Ruin—a useful concept in understanding oil and gas exploration.

BIBLIOGRAPHY

1. Mosteller, Fredrich, et al., Probability and Statistics, Addison-Wesley Publishing Co., Inc., Reading, Mass., 1961; pg. 20.
2. Springer, Clifford H., et al. Statistical Inference, Volume Three of the Mathematics for Management Series, Richard D. Irwin, Inc., Homewood, Illinois, 1966; pg. 131.

7 Binomials and Binomial Coefficients

It's hard to believe that the round-about route of Chapter 6 leads to binomials. Yet, you had to understand permutations before you could comprehend combinations; and combinations are very important to the binomial theorem and binomial expansion.

From binomial expansion stem concepts and knowledge about:

1. risks in multiple well programs (either exploratory or development);
2. the significance of success ratios;
3. the value of multiple exposure; and
4. the number of wells needed to achieve a reasonable degree of success.

The Bionomial

Our next mathematical building block is the binomial. What is a binomial? Bi means two, and nomial stands for number; so, a binomial is an algebraic expression with two terms, e.g.:

$$x + y = 0 \quad \text{or} \quad (a + b)^1 = 0$$

In these expressions, x and y or a and b are the two terms. A binomial expansion is the algebraic expression we get when we multiply a binomial by itself once or more; this is the same as raising it to a power of one or more. For example:

$$(x + y)(x + y) = (x + y)^2 = x^2 + 2xy + y^2 \qquad 7.1$$

Most people remember the last expression automatically, but you can derive it by multiplying as follows:

$$
\begin{array}{r}
x + y \\
x + y \\
\hline
xy + y^2 \\
x^2 + xy \\
\hline
x^2 + 2xy + y^2
\end{array}
$$

$xy + y^2$ ⟵ (multiplying by y)

$x^2 + xy$ ⟵ (multiplying by x)

$x^2 + 2xy + y^2$ ⟵ (summing)

Binomial expansion gets complex when you have powers greater than two. For example, $(a + b)^3$ and $(a + b)^4$ expand to:

$$(a + b)^3 = a^3 + 3a^2 b + 3ab^2 + b^3 \qquad\qquad 7.2$$

$$(a + b)^4 = a^4 + 4a^3 b + 6a^2 b^2 + 4ab^3 + b^4 \qquad\qquad 7.3$$

As you go to larger and larger exponents, binomial expansion becomes very tedious. Calculating the exponents for each number is simple—as one gets smaller the other gets bigger. However, calculating the coefficients, the values in front of each expression, quickly becomes *quite* difficult. Again, mathematicians have figured out a shortcut which reduces the work.

It is the coefficients, the numbers in front of each expression in the binomial expansion, that have our interest; and it just happens that the formula for combinations derives the coefficients. Just why combinations derive the coefficients will be shown later (Equation 7.6). In the preceding expansion of $(a + b)^3$, the coefficients were 1, 3, 3, and 1. To show all coefficients, they are enclosed in parentheses as follows:

$$(1)a^3 + (3)a^2 b + (3)ab^2 + (1)b^3 \qquad\qquad 7.4$$

Let's see if the formula for combinations really does produce these coefficients. (First you need to know that 1! and 0! are both defined as having a value of 1.) The general formula is:

$$_nC_x = \frac{n!}{x!\,(n - x)!} \qquad\qquad 7.5$$

Our coefficients then are:

$$_3C_0 = \frac{3!}{0!\,(3 - 0)!} = \frac{3 \times 2 \times 1}{1 \times 3 \times 2 \times 1} = \frac{1}{1} = 1$$

$$_3C_1 = \frac{3!}{1!\,(3 - 1)!} = \frac{3 \times 2 \times 1}{1 \times 2 \times 1} = \frac{3}{1} = 3$$

$$_3C_2 = \frac{3!}{2!\,(3 - 2)!} = \frac{3 \times 2 \times 1}{2 \times 1 \times 1} = \frac{3}{1} = 3$$

$$_3C_3 = \frac{3!}{3!\,(3 - 3)!} = \frac{3!}{3!} = 1$$

Our answer is 1, 3, 3, and 1; these *are* the coefficients of any binomial expanded to the third power. So the formula for combinations *does* produce the coefficients for binomial expansion.

Now, if we consider n the exponent, then a general equation for the binomial theorem is:

$$(a + b)^n = {}_nC_0 a^n + {}_nC_1 a^{n-1} b + {}_nC_2 a^{n-2} b^2 + {}_nC_3 a^{n-3} b^3 \dots {}_nC_x b^n$$

7.6

By using a slightly different method of display, mathematicians abbreviate this formula as follows:

$$(a + b)^n = \sum_{x=0}^{n} {}_nC_x a^{n-x} b^x$$

7.7

You would read this expression as:

The expansion of $(a + b)$ to the nth power is equal to the sum of the values of all the combinations (from $x =$ zero to $x = n$) of n things taken x at a time, times a to the n minus x, times b to the x.

Note from equations 7.6 and 7.7 that x is always the exponent for (b) the second number of the binomial. Perhaps this relationship will further explain how combinations relate to the coefficients.

This formula is very useful in discussing binomial probability. It is from binomial probability we derive ideas about probabilities for success and failure in exploratory drilling.

Pascal's Triangle

Before moving to binomial probability, however, an important shortcut will be reviewed. A mathematician named Blaise Pascal (1623–1662) was one of the beginning thinkers about probability. He found an extremely simple method of finding the coefficients for the variables in binomial expansion. It is so simple in fact, that you can construct a triangle, called Pascal's triangle, from the binomial expansions just reviewed. We have discussed $(a + b)^1$ for which the coefficients are 1 and 1. For $(a + b)^2$ the coefficients were 1, 2, and 1; for $(a + b)^3$ they were 1, 3, 3 and 1. Arranging these in ascending order:

Binomial			Coefficients			
$(a + b)^0$				1		
$(a + b)^1$			1		1	
$(a + b)^2$		1		2		1
$(a + b)^3$	1		3		3		1

Let's stop here and see what is shaping up. Notice the 2 in the third line of coefficients? It is the sum of $1 + 1$, the numbers to the right and left *above* it. Isn't this also true of the numbers 3 in the fourth line? From this simple beginning you can construct Pascal's triangle—merely by adding the two numbers above to form the one below!

Expanding the previous illustration we have:

TABLE 7.1

PASCAL'S TRIANGLE

Binomial						Coefficients					
$(a + b)^0$						1					
$(a + b)^1$					1		1				
$(a + b)^2$					1	2	1				
$(a + b)^3$				1	3		3	1			
$(a + b)^4$			1	4		6		4	1		
$(a + b)^5$			1	5	10		10	5	1		
$(a + b)^6$		1	6	15		20		15	6	1	
$(a + b)^7$	1	7	21	35		35		21	7	1	
$(a + b)^8$	1	8	28	56	70		56	28	8	1	
$(a + b)^9$	1	9	36	84	126		126	84	36	9	1
$(a + b)^{10}$	1	10	45	120	210	252	210	120	45	10	1

You could carry this experiment out as far as you like. What marvelous time-savers come from shortcuts! Pascal really found a valuable shortcut to values for $_nC_x$. We shall come back to this table several times to take advantage of the shortcut.

Just to refresh your memory as to the meaning and significance of Pascal's triangle an illustration follows. Check the last line of Table 7.1. Here n equals 10, and, for the sixth figure in that line, x equals 5; so the coefficient should be $_{10}C_5$. The proof is:

$$_{10}C_5 = \frac{10!}{5!(10 - 5)!} = \frac{10 \times 9 \times 8 \times 7 \times 6 \times 5!}{5!(5!)}$$

$$= \frac{10 \times 9 \times 8 \times 7 \times 6}{5!} = 252$$

The future use of $_nC_x$ will involve concepts of *drilling* and *success*—where n will be the number of wildcats drilled and x the number of successes (or discoveries).

Before leaving Pascal's triangle another arrangement deserves a review. The figures from Pascal's triangle can be arranged "on their side," so to speak, to illustrate another point. For example—

$(a + b)^0$	1					
$(a + b)^1$	1	1				
$(a + b)^2$	1	2	1			
$(a + b)^3$	1	3	3	1		
$(a + b)^4$	1	4	6	4	1	
$(a + b)^5$	1	5	10	10	5	1
etc.						

Note the second column in the new arrangement. It represents the exponent in the binomial expansion. If you refer to Table 7.1 you will see that the column above is equivalent to the second diagonal column in Pascal's triangle. Now you can easily locate the exponent of the binomial expansion.

We next move from binomial expansion to binomial probability.

Binomial Probability

In making the change from *binomial expansion* to binomial probability, it is necessary to change the notation of the things in the following manner:

Binomial Expansion		Binomial Probability	Meaning of the Term
a	becomes	q	probability of failure
b	becomes	p	probability of success

So, $(a + b)^n$ becomes $(q + p)^n$—or preferably $(p + q)^n$. If you come across binomial probabilities elsewhere, the $(p + q)^n$ notation is what you will most likely find.

If p is the probability of an event occurring in any single trial (you can call this the probability of *success*) and $q = 1 - p$ (the probability of a dry hole)—then the probability that the event will happen exactly x times in n trials (i.e., x discoveries and $n - x$ dry holes) is given by:

$$p(x) = {}_nC_x p^x q^{n-x} = \frac{n!}{x!\,(n - x)!}\, p^x q^{n-x} \qquad 7.8$$

This equation is called the *binomial probability function*. Note how similar it is to Equation 7.7. Now you can see the value of all of our work to this point. The lengthy discussion of binomial expansion and binomial probability has been the ground work necessary to get to this one important equation!

You can substitute D and S in the binomial probability formula (7.8) to indicate the probability of a dry hole or a successful well:

$$p(x) = {}_nC_x S^x D^{n-x} \qquad 7.9$$

In formula 7.9:

   $p(x)$ = probability of *exactly* x successes
   x = number of successful wells
   n = total number of wells
   S = probability of success
   D = $(1 - S)$ or probability of each well being dry
   ${}_nC_x$ = a coefficient which is our familiar formula for *combinations*—or the number of combinations of n things taken x at a time.

Before using this newly found formula, let's go back to a simple illustration.

Suppose you are dealing with several things which might happen and want to relate these to what will happen. One might say this as:

Lots of things *can* happen but only one thing *will* happen. A mathematician would say that an equation for this idea would read:

The sum of the probabilities of all the things that can happen must equal what does happen (which must be one of the things which can happen),

or more succinctly:

The sum of the probabilities of all possible outcomes = 1.0.

Let's now bring this knowledge to bear on an example involving wildcat wells. As shown in Equation 7.9 we will use symbols D and S to indicate:

   D = dry hole
   S = successful well

An easy way to begin is with a three well program. This example could, however, also be used to illustrate the possible outcomes in a development well with three potential pay zones.

In drilling only three wells there are eight possible sequences of events which can occur. A mathematician would state this relationship as 2^3 or $2 \times 2 \times 2$ or 8. These eight outcomes are shown in Table 7.2.

TABLE 7.2

ALL POSSIBILITIES IN A 3 WELL PROGRAM

D = Dry	S = Success
DDD	All 3 Dry
DDS	2 Dry—3rd Successful
DSD	2 Dry—2nd Successful
SDD	2 Dry—1st Successful
SSD	2 Successful—3rd Dry
SDS	2 Successful—2nd Dry
DSS	2 Successful—1st Dry
SSS	All 3 Successful

The outcomes range from all dry to all successful. Three possible combinations include two dry holes and one success; and three combinations could produce only one dry hole and two successes.

If we assume that D and S are *equally* probable, then this simple arrangement shows us that each event has a chance of occurring once in eight times; or we could say each event has a probability of 0.125. However, there are three events which produce one discovery; therefore, the chance for one discovery (if D and S are equally probable) is three in eight or 3/8. We would say one discovery has a probability of *0.375*. The same probability of 0.375 applies to two discoveries since there are also three events which could produce that result. Only one chance in eight exists for three successes—the same as three dry holes.

The Outcomes from Binomial Expansion

We will now proceed to use our knowledge of binomial expansion to bear on this illustration. We can express in binomial expansion the same illustration shown in Table 7.2. The eight outcomes just discussed for our three well program can be expressed algebraically as shown in Table 7.3.

TABLE 7.3

BINOMIAL EXPANSION

$$(D + S)^3 = 1$$

D^3	$+ 3D^2 S$	$+ 3DS^2$	$+ S^3$	$= 1$
DDD	DDS	SSD	SSS	
	DSD	SDS		
	SDD	DSS		
1	3	3	1	

Note the expansion of $(D + S)^3$. It yields the same set of outcomes listed in Table 7.2. This time, however, they have been derived from binomial expansion. The eight possible outcomes are listed below their appropriate algebraic expression. Here again we see the higher probabilities (if D and S are equally probable) of one or two successes.

The expression $(D + S)^3$ was set equal to one. We do this to acknowledge the fact that only one event can occur and that the sum of the probabilities of all possible outcomes must equal one.

Note also that the sum of the coefficients equals the total number of events; i.e., $1 + 3 + 3 + 1 = 8$. Remember these numbers from Pascal's triangle? We now have another important fact to use later. It is:

"the sum of the coefficient in Pascal's triangle for a specific binomial expansion equals the total number of possible events."

Here's how we put this relationship to use. We now know that if D and S are equally probable, the chance of all three wells being successful (SSS) is only one in eight, or

$$\frac{1}{1 + 3 + 3 + 1} = .125$$

We could develop this same data from the binomial probability function as follows:

$$p \text{ (for 3 successes)} = {}_3C_3 \, (0.5)^3 \, (0.5)^{(3-3)} \qquad \qquad 7.10$$

$$= {}_3C_3 \, (0.5)^3 \, (0.5)^0 = {}_3C_3 \, (0.5)^3$$

$$= \frac{3!}{3! \, (3-3)!} \, (0.5)^3 = 1 \, (0.5)^3$$

$$= (0.5)(0.5)(0.5) = 0.125$$

Perhaps now you can see why Pascal's triangle provides such a convenient and time-saving shortcut, if D and S are equally probable. Although the outcomes from the two events (D and S) can be determined by the equation, it becomes more cumbersome as the number of wells (n) in a drilling program increases. A four well program has (2^4) events or 16. A five well program has (2^5) or 32. (Remember we can also get the number from Pascal's triangle—for a five well program $1 + 5 + 10 + 10 + 5 + 1 = 32$.) So as the number of wells drilled increases we have more difficulty in easily getting at the probabilities for specific successes—unless we can use the coefficients in Pascal's triangle.

Some authors use different symbols for the prefix of the binomial probability function (Eq. 7.9). One such prefix starts with b (or B) indicating binomial probability function followed by the three key factors which are:

x = the number of successes
n = number of trials
p = the probability of success.

The prefix of the equation thus becomes $B(x, n, p)$—and is read as "the probability of getting exactly x successes from n trials given a chance of p for one trial." Using this format we could rewrite our prior illustration as:

$$B(3, 3, 0.5) = {}_3C_3 (0.5)^3 (0.5)^{(3-3)} = 0.125 \qquad 7.11$$

This expression is identical to Equation 7.10 except that our prefix gives us the three key parameters of x, n and p.

As is so often the case in exploring for oil and gas fields, D and S do not have an equal chance of occurrence. What happens in our three trials to the probability of three successes if the chance of success is not 50% but only 20%? Using the format of Equation 7.11, we have:

$$B(3, 3, 0.2) = {}_3C_3 (0.2)^3 (0.8)^{3-3} \qquad 7.12$$

$$= (0.2)(0.2)(0.2) = .008$$

If the chance of success is only 20%, then the probability of three successes in only three trials is one in 125 or 0.008. Remember when p is 0.5 (50% chance) the probability of three successes in three trials is one in eight (0.125).

You should be off and running now—you can use this new formula for any set of conditions. Before you do, let's work another example.

Given:

We will have a 10 well program $n = 10$
The success rate has averaged 30% $p = 0.3$

Question:

What are our chances of exactly two successes in 10 wells with an average success rate of 30%?

Answer:

$$B(x, n, p) = {}_nC_x p^x q^{(n-x)} \qquad 7.13$$

$$B(2, 10, 0.3) = {}_{10}C_2 (0.3)^2 (0.7)^8$$

$$= \frac{10!}{2!\,(10-2)!}\,(0.09)(0.0576)$$

$$= \frac{10 \times 9 \times 8!}{2!\,8!} \times (0.09)(0.0576)$$

$$= 45 \times .005188$$

$$= .233$$

The probability of exactly two successes in 10 trials, where the chance of success is 0.3, is only 0.233. Let's work one more example.

$$B(3, 7, 0.4) = \frac{7!}{3!\,(7-3)!} = (0.4)^3 (0.6)^{(7-3)} \qquad 7.14$$

$$= \frac{7 \times 6 \times 5 \times 4!}{3! \times 4!}\,(0.4)^3 (0.6)^4$$

$$= \frac{210}{6}\,(.064)(.1296) = 35 \times (.00829)$$

$$= 0.290$$

For exactly three discoveries in seven wildcat wells in a trend where our chance of success is 40%, the probability is 0.290.

Cumulative Probabilities

Now let's pose a slightly different problem. Suppose someone asks you the probability of getting 4 *or more* heads in six tosses of a coin. Remember in coin tossing p is always 0.5. The question posed differs substantially from the chance of exactly four heads in six tosses. To answer 4 *or more* you must consider the probabilities of getting 4 heads, 5 heads and 6 heads—because each condition would satisfy four or more. Now the formula takes a different shape:

$$B(4 \text{ or more}, 6, 0.5) = B(4, 6, 0.5) + B(5, 6, 0.5) + \qquad 7.15$$

$$B(6, 6, 0.5)$$

$$B \text{ (4 or more, 6, 0.5)} = {}_6C_4(0.5)^4 (0.5)^2 + {}_6C_5(0.5)^5 (0.5)^{6-5}$$
$$+ {}_6C_6(0.5)^6 = 0.344$$
$$B \text{ (4 or more, 6, 0.5)} = .234 + .094 + .016 = .344$$

The probability of 4 or more heads in six tosses is 34.4%. The expression "four or more" is mathematically the same as saying "at least four."

To refresh your memory let's look at the coefficients from the expansion of $(a + b)^6$ (Table 7.1). We shall list them and then state what they stand for in terms of success or dry.

Coefficient	Meaning
1	All dry
6	five dry—one success
15	four dry—2 successful
20	3 dry—3 successful
15	2 dry—4 successful
6	1 dry—5 successful
1	all successful
64	Total possibilities

Remember, these coefficients represent the number of possibilities for each meaning; they can be used directly if $p = 0.5$ (D and S are equally likely). Thus, B (4 or more, 6, 0.5) can be calculated as follows from the coefficients from Pascal's triangle:

$$B \text{ (4 or more, 6, 0.5)} = \frac{15}{64} + \frac{6}{64} + \frac{1}{64}$$
$$= .324 + .094 + .016 = .344$$

The answer is produced again by a different method!

You can see how cumbersome it would be to calculate the binomial probability of an outcome for a large number of trials. It is even worse to calculate the cumulative binomial probability (*x or more* successes). Fortunately another shortcut exists in the form of tables from which one can read probabilities directly. Using tables is much faster than having a large Pascal's triangle for binomial coefficients ($_nC_x$); and the tables require no calculations—just an understanding of how they are constructed.

Table 7.4 contains an excerpt from such tables. Note it lists *cumulative* not individual probabilities. That means the probabilities are for *x or more*, not just *x*. Tables have been constructed for both B (*x, n, p*)

TABLE 7.4

CUMULATIVE BINOMIAL PROBABILITY
$B(x$ OR MORE, n, 0.2)

n	Values of x						
	1	2	3	4	5	6	7
2	.36	.04					
3	.49	.10					
4	.59	.18	.03				
5	.65	.26	.06				
6	.74	.34	.10	.02			
7	.79	.42	.15	.03			
8	.83	.50	.20	.06	.01		
9	.87	.56	.26	.09	.02		
10	**.89**	.62	.32	.12	.03	.01	

and B (x or more, n, p). In Table 7.4 the data are listed for the conditions where $p = 0.2$—20% chance of success. Values of x go to 7 and values of n are shown to 10. In Appendix C are several tables of both individual and cumulative probabilities.

To use tables such as 7.4 locate the number of trials (n) and the corresponding values of x or more and listed to the right of the value of n. For example, the probability of one or more successes in ten trials is 0.89—for 4 or more successes in 10 trials, 0.12, etc.

An Application

A brief example on wildcat drilling can illustrate the useful insights to be gained from binomial expansion. You have often heard that many companies and individuals try to spread their risks in trends or plays where little is known and the money involved is large. Just why is a one-half interest in two wells less risky than a full interest in one well? Why is a one-fourth interest in four wells less risky than a full interest in one well?

You may feel intuitively that these questions are answered by inspection. Binomial expansion, however, can show us exactly why individuals may desire to spread the risk.

TABLE 7.5

REDUCING THE RISK WITH MULTIPLE WELLS

One-Well Program

Outcome	Probability	AVP(L)—M$	Risk-Weighted AVP—M$
D	.9	−100	−90.0
S	.1	1,000	100.0
	1.0		10.0

Two-Well Program

Outcome	Probability	AVP(L)—M$	Risk-Weighted AVP—M$
DD	.81	−200	−162.0
DS	.09	900	81.0
SD	.09	900	81.0
SS	.01	2,000	20.0
	1.00		20.0
			One-half interest in 2 wells = 10.0

Three-Well Program

Outcome	Probability	AVP(L)—M$	Risk-Weighted AVP—M$
DDD	.729	−300	−218.7
DDS	.081	800	64.8
DSD	.081	800	64.8
SDD	.081	800	64.8
SSD	.009	1,900	17.1
SDS	.009	1,900	17.1
DSS	.009	1,900	17.1
SSS	.001	3,000	3.0
	1.000		30.0
			One-third interest in 3 wells = 10.0

The demonstration of the answer is shown on Table 7.5. It uses the payoff table concept to illustrate monetary gain and exploratory risk. A one well, two well and a three well program are illustrated.

The assumptions for the example are:

a. Only 10% chance for a successful wildcat
b. A large inventory of prospects, similar in size and risk.

The payoff tables show:

1. The possible outcomes from each well program.
2. The probability of their occurrence.
3. The economic reward (actual value profit—AVP)—it can also be an actual value *loss*.
4. The risk-weighted profit which is the actual value profit times its probability of occurrence.
5. The expected value which is the sum of the risk-weighted values for all possible outcomes.

What does our example show? In the one well program the expected value for two possible outcomes is $10,000. The dry hole, which will occur 9 out of 10 times, represents a risk-weighted loss of $90,000. A successful well would produce $1,000,000; but since it will occur only once in ten wells, the risk-weighted profit is only $100,000. Thus, the expected value, the sum of the two risk-weighted values, is $10,000.

The two well and three well programs show that one-half of the expected value in a two well program and one-third of the expected value in a three well program are also $10,000. The monetary gain *remains the same* under the conditions set forth!

What does change? The probability of all wells being dry changes! For a one well program the chance of it being dry is 90%. For a two well program the chance of all being dry is 81%; and for a three well program the chance of all dry is 72.9%. The chance of losing everything *decreases* with a lower participation in a larger number of wells. In this way risk is *reduced*.

A firm with a strong risk aversion willingly sacrifices the likelihood of above average outcomes for a reduction in the likelihood of below average chances of losing everything.

One other factor does also change. The more wells in which you have an interest the better your chances of getting one of the larger fields in a lognormal distribution. This condition was not allowed in our original set of conditions as we assumed prospects of similar size and risk. Large fields are rare; and so the only way you increase your chances of participating in the larger finds—in a statistical sense—is to participate in a greater number of exploratory wells.

Mathematically the preceding statement can be illustrated by calculating the mean of an increasingly larger sample of fields from the *same* lognormal distribution. Table 7.6 shows 5, 10, and 20 field samples from the lognormal distribution in Fig. 7.1. Here the assumption is made

TABLE 7.6

MEAN INCREASES WITH NUMBER OF FIELDS

	5 Fields		10 Fields		20 Fields	
No.	Value*	Midpoint	Value*	Midpoint	Value*	Midpoint
1	27.0	10	50.0	5	85.0	2.5
2	7.6	30	18.0	15	35.0	7.5
3	3.2	50	10.0	25	22.0	12.5
4	1.3	70	6.0	35	15.0	17.5
5	0.4	90	3.9	45	11.3	22.5
6			2.6	55	8.6	27.5
7			1.7	65	6.5	32.5
8			1.0	75	5.4	37.5
9			0.5	85	4.3	42.5
10			0.2	95	3.5	47.5
11					2.8	52.5
12					2.3	57.5
13					1.8	62.5
14					1.4	67.5
15					1.1	72.5
16					0.9	77.5
17					0.7	82.5
18					0.5	87.5
19					0.3	92.5
20					0.2	97.5
Total	39.5		93.9		208.6	
Mean	7.9		9.4		10.4	
Median	3.2		3.2		3.2	

*$\bar{M}$ bbl.

that all are fields which will align themselves along the line of the parent distribution; i.e., the five field sample was read from the curve of Fig. 7.1 at the midpoints indicated; the ten fields were read from the midpoints shown as were the 20 fields. Thus, each sample of fields was spaced equidistant (percentagewise) along the one lognormal curve.

Note how the mean values of the distributions increase with field sample. The five fields have a mean of 7.9 million bbl; the ten fields a mean of 9.4 million and the 20 fields have a mean of 10.4 million bbl. The median, however, remains the same. The incremental increase in mean size gets smaller and smaller as the number of fields increases. You can see how rapidly this occurs by the curve in Fig. 7.2. Note also how critical the curve is up to about the sixth discovery. Since from 20 to 50 wildcats would have to be drilled to get five to ten discoveries, the sharply breaking early part of the curve is the most critical.

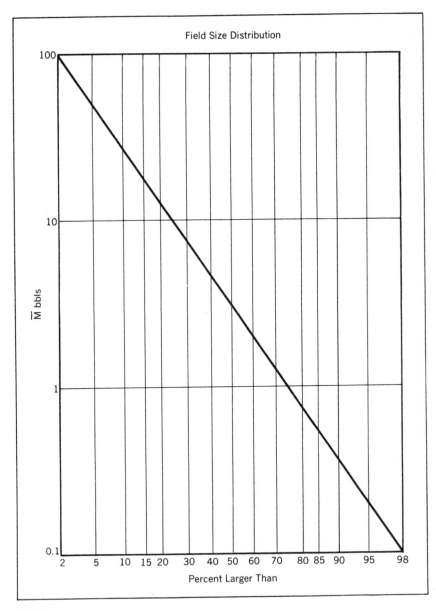

FIG. 7.1

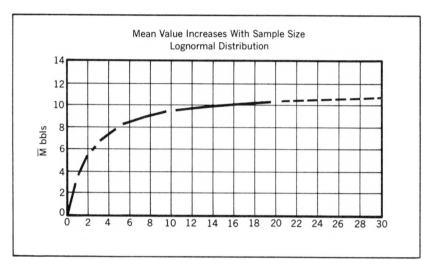

FIG. 7.2

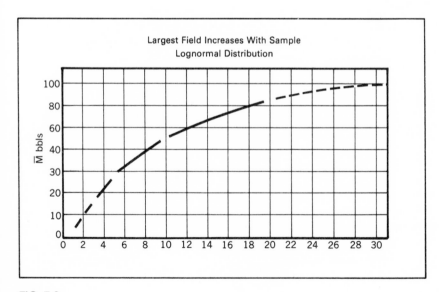

FIG. 7.3

Just precisely why the mean increases can be seen from Fig. 7.3 which shows the increase in the size of the largest field with sample size. Again the curve rapidly decreases in slope after the 20th discovery.

What all of these relationships reveal is another of the practical values of binomial expansion. Spreading the risk enhances your possibilities of getting a larger mean reserve; and as we have seen, it reduces your chance of them all being dry!

REVIEW

Following an introduction to permutations and combinations, this chapter introduced the binomial theorem, binomial expansion and Pascal's triangle. Any two event system benefits from knowledge about binomial theorem. The exploratory well with its outcomes of dry or successful fits into a two event system. Probabilities of success or failure can be derived for any level of drilling from binomial theorem. Pascal's triangle gives shortcuts to these calculations for the special case where the probabilities of the two events are equally likely. For success rates (values of p) other than 50%, tables from statistical reference books can be used to determine probabilities of outcomes.

Binomial probabilities also help illustrate the benefits of spreading risks. The next chapter will show how binomial probabilities can help in determining the number of wells needed to achieve a desired degree of success.

Part II
APPLICATIONS

The next chapter begins Part II. These chapters build upon the base of the first seven chapters. They assume you understand the importance of lognormality and the significance of binomial theory to the two event system of drilling for oil and gas. These chapters will provide simple illustrations of risk analysis concepts, some of which are recommended and some are not. Those concepts not recommended are included only because you will run across them in the literature. You need to know their limitations.

Wherever possible new concepts are explained in non-mathematical terms. In a few chapters, because of the proprietary nature of the state of the art, only the available published literature is reviewed. However, the important conclusions about risk analysis are almost all available from the literature. Risk analysis has been around a long time. What has changed is that more and more explorationists are becoming aware of its value to their exploration programs. To a large degree the sealed bid sales in the outer continental shelf have done more than anything else to advance the use of risk analysis. The limited geological knowledge and high monetary commitments have necessitated the consideration of every possible useful technique.

8 "Gambler's Ruin"

Searching for hydrocarbons may well be the world's biggest gambling game. The stakes are high and the results are uncertain. Nature represents the house and she wins much of the time. In gambling the odds favor the house. It is as if the house is drilling development wells with its money and your money goes for wildcats. Even in the simple toss of a coin (which has the high probability of success of 1/2), you can have a long run of either heads or tails; and if the stakes are high and your capital is minimal you can be ruined.

It is for this reason that gamblers have toyed with a fundamental concept called Gambler's Ruin. This rule, based on binomial theory, says that the higher the risk (that is, where the chances of success are small) the greater the chance of going broke from a normal run of bad luck. The concept holds regardless of long run expectations of a profit!

There is no way to avoid this risk absolutely; but you can reduce it by making sure that your exposure, when related to your capital, has only a small chance of loss. This really means you need enough capital to ride out the inevitable runs of bad luck. When a business fails due to Gambler's Ruin, we say it was "undercapitalized."

Normal Runs

What are "normal" runs of bad luck? How do we determine them? How does this knowledge help us? The answer to the first question can never be answered with absolute exactness. Yet normal runs of "bad luck" can be inferred from binomial probabilities; and that's what we've been talking about.

Concepts involving Gambler's Ruin have some distinct limitations. These will be discussed. Nevertheless, the term "Gambler's Ruin" appears in the literature frequently and a few authors still espouse its usage.[1] Thus even with its limitations, it needs discussion.

In a figurative sense, an insurance company "places so many bets" that it is not affected by "Gambler's Ruin."[2] But an oil company does not have the luxury of thousands of bets (wells). Therefore, information from "Gambler's Ruin" does have value. For example, if you drill ten wells in an area where the success rate has averaged 10%, the law of "Gambler's Ruin" would tell you that you had a 35% chance that all ten wells would be dry! The purpose in this chapter is to show you the exact basis for the preceeding statement.

Graphing Cumulative Binomial Probability

We saw in Table 7.4 that tables exist which display cumulative binomial probabilities—showing the probabilities for various successes (x or more) for varying numbers of wells (n). In Table 7.4 the heavy blocked number was the answer to:

$$B(1 \text{ or more}, 10, 0.2) = P(1) + P(2) + P(3) + P(4) \dots P(10)$$

$$= .27 + .30 + .20 + .07 \dots$$

$$= 0.89$$

The probability of one or more (sometimes read as "at least one") discoveries for ten wells in a trend with 20% success rate is 0.89 or 89%. Note that the highest individual probability (0.3) is nearest the average probability ($P = 0.2$) or at exactly two successes. A review of the nature of the formula for binomial probability will show you that this will always be true. The series $P(1) + P(2) + P(3)$ etc. is not a progression numerically downward from $P(1)$.

On Fig. 8.1 a plot of cumulative binomial probabilities is shown for the special case where $P = 0.2$ (20% chance of success). The top curve is labelled "at least one." It can be shown also as $\leq$ which means "equal to or less than." Thus, the top line in Fig. 8.1 shows the case of one or more discoveries at various numbers of wildcats (values of n)—these values being the horizontal axis. Note the intersection of the top curve on the line for $n = 5$. This intersection "says" that if $n = 5$ there is a 65% chance of at least one discovery. For 10 wells there is an 89% chance of at least one discovery. Thus we can show graphically the numbers contained on tables of cumulative binomial probability.

The Case of $X = 0$

There is a special case of individual probability which has a direct bearing on concepts of "Gambler's Ruin." That case is the one where $x = 0$. It represents the situation where all wells were dry—none

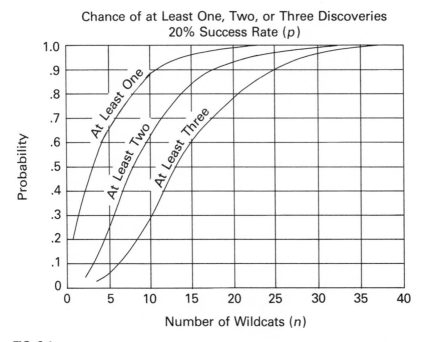

FIG. 8.1

successful. The reason this special case occupies our interest for "Gambler's Ruin" is this: the basic question answered by Gambler's Ruin is—"What is the probability that all wells are dry?". The effective use of "Gambler's Ruin" concepts comes when you utilize the answer to this question to make investments such that potential losses are minimized.

Fig. 8.2 was generated by plotting from tables of individual binomial probabilities the values for $x = 0$ for various values of n. In the figure "n" is the horizontal axis and probability is the vertical axis. The probability that all wildcats will be dry is shown for three success rates (3 values of p)—namely 10%, 20%, and 30%.

First let's examine the 20% line. Remember from Table 7.4 and Fig. 8.1 the chance of *at least one* discovery was 89%. Therefore, the chance that ALL would be dry is one *minus* the chance of at least one discovery.

$$B(0, 10, .1) = 1 - .89 = 0.11 \text{ or } 11\%$$

Now read the 20% line on Fig. 8.2 at the value of $n = 10$. You see

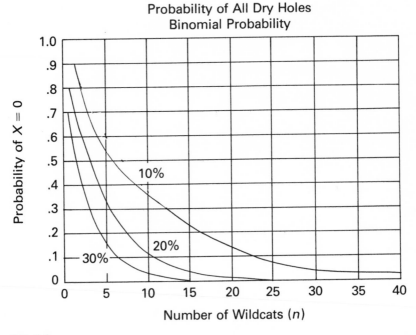

Probability of All Dry Holes
Binomial Probability

FIG. 8.2

that it shows a chance of 11% for all wells being dry. The 30% line shows only a 3% chance of all dry at $n = 10$; the 10% line shows a 35% chance of all ten wells being dry (if $p = 0.1$).

Remember the statement made in the sixth paragraph of this chapter. You have a 35% chance of all dry holes in a ten well program where the success rate averages *10%!* Now you see where that statement came from—i.e., exactly how it was derived.

Using Concepts Derived from $X = 0$

The Law of "Gambler's Ruin" is one extension of the use of curves or data from individual probabilities of $x = 0$. There are very simple uses of the concept; and there are more complex usages. A recent one referred to earlier was by Arps and Arps in the July 1974 issue of the Journal of Petroleum Technology. Their article was entitled "Prudent Risk-Taking." The authors stated that prudent risk-taking involved two concepts:

1. Successive risk-taking ventures should be undertaken only under a break-even situation in the long term.
2. The money risked on a single venture should not exceed that which would increase the risk of "Gambler's Ruin" beyond *acceptable limits.*

Note the term "acceptable limits" in the second concept. The investor must determine *for himself* what his acceptable limits are. The maximum would be the limits for *total loss.*

An Example

Let's now use a very simple example to illustrate these concepts. The assumptions are:

1. There are ten prospects.
2. The drilling capital is $1,000,000.
3. Dry cost per wildcat well is $100,000.
4. Expected average success rate is 20%.
5. A 95% assurance of at least one discovery is desired.

Given these facts as input we seek answers to the following questions:

1. What are the chances of all ten prospects being dry?
2. With ten prospects is there 95% assurance of one or more discoveries?
3. If not, how much larger must "n" be (how many more wells), or
4. With 10 wells what will the success rate be to have 95% chance of at least one discovery?

Now to review each question, beginning with the first.

The answer to question one has already been provided from a prior illustration. When $p = .2$ and $n = 10$, there is still an 11% chance that all wells would be dry. This answer comes from the 20% chance curve in Fig. 8.2.

If there is an eleven percent chance of all being dry, then obviously the answer to question two is "no". We cannot be 95% sure of at least one discovery. Thus, if 95% is the "acceptable limit" we would reject this series of prospects.

On the other hand, Fig. 8.2 shows that if 14 wells are drilled (instead of 10) there is a 95% chance of at least one discovery. The answer to question three is 14 wells.

However, there is only enough capital to drill 10 wells—not 14. Suppose the estimate of *average* success is conservative. How much would success rate have to be to give the 95% assurance of one or more discoveries? The answer to this question requires interpolation between the 20% and 30% curves in Fig. 8.2. A success rate of approximately 26% is needed to have a 95% chance of one or more discoveries with only 10 wells.

Notice we have only dealt with the chance of success, not the chance of profit. Profit must be worked into the solution of this problem. A small discovery might not return the $1,000,000 of invested capital. The assumption here is that all ten wells are drilled. One possibility is that the first two or three holes could be dry and could also prove false the basic geologic assumptions upon which the prospects were isolated. You could stop drilling then. On the other hand, that first well could be a discovery and thus finance the whole play.

A Normal Run of "Bad Luck"

One important point needs special comment. If you drill 14 wells, are you *guaranteed* at least one discovery in our example above? The answer is NO! There is no guarantee. Binomial probability is based on long term averages but short term variations are possible. Remember the coin tosses in Chapter 2. Even 100 tosses did not reproduce the long term average frequency of heads.

The values of $x = 0$ are mathematical expressions of a "normal run of bad luck" in a two event system. "Normal", then, is the long-term expected average from many "coin tossings" or many well programs. In the example of 10 and 14 wells it is still possible to drill even 20 wells all dry, but if $p = .2$ the probability is low. However, even low probabilities can and do occur. If it can happen, it will happen sometime! Even in a 10,000 to one odds, it could happen the first time!

Another Example

Before discussing some limitations of binomial probability let's illustrate another simple example:

Given:

Capital of $5,000,000
Dry well cost $250,000
Acceptable limits 90% chance of at least one discovery

Solution:

Success Rate (p)	Prospects (n) for Only 10% Dry Chance
10	23
20	11
30	7

Without posing questions, the table tells us:

1. 23 prospects are required to satisfy the acceptable limits at 10% success rate.
2. The capital is not sufficient for this type of play.
3. At 20% success rate only 11 prospects satisfy the acceptable limits.
4. At 30% only 7 prospects are needed.

These answers were derived from binomial probability tables as in the previous example. From these simple examples you can see the types of applications made from the "law of Gambler's Ruin." However, this concept has some severe limitations which must be considered in its use.

Limitations

1. First of all, not all runs of bad luck are "normal." The estimate of the number of wells it takes to achieve acceptable limits is itself an average based on the results of many events.
2. Secondly, the probabilities are assumed constant with time; i.e., they do not change as we drill each prospect. This assumption doesn't fit most plays where each new well adds significant data.
3. Third, binomial theorem assumes each event is independent of other events—which may not be the case. If one well is related to the outcome of another the events are *not* independent.
4. The "Law of Gambler's Ruin" has no application to a single event (one prospect). This realization reaffirms the geologic common sense of trying to higrade prospects—to drill the best ones first!

5. For exploration, binomial theorem assumes a constant supply of prospects—sampling *with* replacement, a mathematician would say. However, the population diminishes with each new discovery. Newly discovered fields are not replaced but represent one less left to find.

6. Finally, success is not quantified. The size of discoveries is ignored, and this factor must enter your decision process. The average success rate is related to the chance of getting *anything*, the minimum size or more. At least one discovery doesn't guarantee a field large enough to make a profit.

The Uses

With all these limitations, of what value is the "Law of Gambler's Ruin?" What good is it to know what constitutes a normal run of bad luck?

1. "Gambler's Ruin" concepts re-emphasize important aspects about success ratios. They remind us that success ratios are averages based on many events.

2. Because success ratios are long term averages, any small group of prospects can vary considerably from that average.

3. "Gambler's Ruin" shows how far a given number of wells would "normally" vary from the average. It does this by illustrating a normal run of bad luck.

4. Concepts about "Gambler's Ruin" can foster an understanding of the probabilities associated with exploration drilling, development drilling, and multiple pay field wells.

5. The dimensions of a particular deal can be expressed in the number of wells or success ratios needed to achieve a desired goal.

6. Concepts of "Gambler's Ruin" can demonstrate for the non-explorationist that in a 10 well program, a 20% success rate does *not* automatically mean *two* discoveries—even though that is the most probable outcome!

When the risks are great, there is no absolute rule; but any insight which helps us more fully understand an opportunity is of value. It is in this sense that binomial theorem and "Gambler's Ruin" have value and meaning to the explorationist.

REVIEW

A concept of the chances of ruin has meaning to a company with limited capital. The meaning is general and philosophical in nature. The geology of a prospect still is its most important aspect. For the company, not capital limited, understanding "Gambler's Ruin" adds a philosophical dimension to the array of data about an investment. Even after considering the geology, geophysics and economics, you still need to consider the probabilities of a completely unsuccessful play.

An understanding of "Gambler's Ruin" can help explain some of the mysteries others see in the petroleum industry. Never has there been a more important time to help the public understand the industry. A little math, a table or a chart may help explain the difficulties faced even under the best of nature's circumstances.

BIBLIOGRAPHY

1. Arps, J. J. and Arps, J. L., "Prudent Risk-Taking," Journal of Petroleum Technology, July, 1974, pg. 711.
2. Slichter, L. B., "The Need of a New Philosophy of Prospecting," Mining Engineering, June 1960, pg. 570.

9 Opinion Analysis
or
The Case for
Subjective Probability

"Even when all the experts agree, they may well be mistaken."
— Bertrand Russell

". . . . the greatest error in forecasting is not realizing how important are the probabilities of events other than those everyone is agreeing upon."
—Paul Samuelson
Bus. Week 12/21/74, pg. 51

"The man who insists upon seeing with perfect clearness before deciding, never decides."
—Henri Fredric Amiel

Almost every human activity involves decisions which must be made where proper information for assuring the right decision is lacking. These activities can be global or individual. On a global basis, mankind perpetually seeks solutions to famine, inflation, resource shortages, and war. The uncertainties here seem obvious; but there are equally difficult decisions to be made in an individual human life. Choosing between job opportunities, buying an automobile, or even selecting a spouse entails decisions which often have an inadequate base of data.

What Is Probability?

Whenever the number of possible outcomes for an event is numerous we can only surmise as to what is probable. Some events can have their probabilities depicted in distributions as illustrated in Chapter four. In that chapter we showed that some distributions are discrete; i.e., only a certain, limited number of outcomes is possible. Rolling two dice represents a discrete distribution. Only 36 possible combinations of two dice can occur. Furthermore, some outcomes can be shown to be more probable than others. A distribution with an unknown number of possible outcomes is commonly called a continuous or non-discrete distribution. Many situations exist where the number of possible outcomes is not known. Obviously in these instances the data base is obscure. Probability and probability concepts are attempts to assign the frequency of possible happenings to the happenings. There are two types of probability.

Objective probability is based on either a discrete set of outcomes or upon data obtained from reasonable, empirical experience.

Subjective probability must be utilized when all of the outcomes are not known and when even the frequency of those outcomes recognized cannot be estimated with certainty. Objective probability helps in understanding the choices in gambling. Subjective probability is what explorationists use in the search for oil and gas fields. For the explorationists, neither the number of fields to be found is known, nor is their size. Furthermore, the explorationist has no guarantee, for a small number of wildcats (a small sample), of an average success rate. He may not think of it as subjective probability but the explorationist uses his past experience and judgment to make decisions about future events. He uses subjective probability or as some persons like to think of it—opinion analysis.

The Knowledge Curve

The relationship between subjective probability and objective probability is shown by Fig. 9.1. Knowledge or information is plotted against the range of opinion. When knowledge is large (when the facts are many) factual analysis is possible. We can use objective probability. However, when knowledge is limited (when the facts are few) we must rely upon an analysis of opinion or subjective probability.

The knowledge curve shows an important fact about subjective probability. When the facts are few, the range of opinion is broad. Therefore, *expect* broad differences of opinion in risk analysis.

Important References in Recent Literature

Within the last decade an increasing number of writers have sought to explain and espouse the value of subjective probability in risk analysis. Several of these references will be reviewed in the paragraphs to follow.

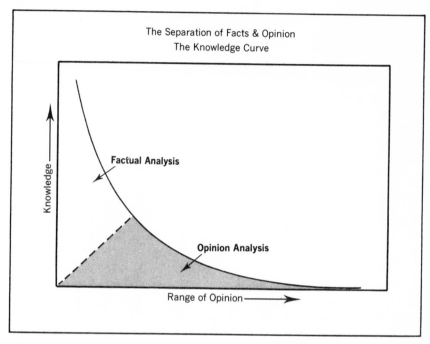

FIG. 9.1

Hertz

One particular paper is a constant reference in most articles or books dealing with risk analysis. That paper, published in a 1964 issue of the Harvard Business Review, was written by David B. Hertz.[1] Much of the substance of the article was later published in a portion of a book on management science.[2]

Hertz's initial article is a classic. Its succinct articulation of an approach to dealing with uncertainty—as well as its lucid clarity—probably account for its popularity. Many managers would be utilizing subjective probability if the literature contained more articles with the simplicity and ease of understanding present in Hertz's HBR article.

Hertz began that article with a discussion of the impact of assumptions on any answer in capital investments. The printed page from computer generated calculations looks so deceivingly formal and precise. Yet, behind the printed page lie data which are anything but precise. The calculation of investment yardsticks (rate of return, profit, payout, etc.) involves numerous variables, each of which has a large degree of uncertainty. Most managers recognize that assumptions determine an answer; and that the answer is no better than the accuracy of its underlying assumptions. Nevertheless, there are managers who, faced with vast uncertainty in basic assumptions, still insist on a single answer.

Hertz illustrates the fallacy of a "most likely" answer with the use of 5 variables. Each variable has only a 60% chance of being right. Thus the most likely probability would be 0.6 (60%) for each variable. However the chance that each variable will be exactly right to produce the most likely answer is very low—in fact only 8%! This number is calculated by .6 × .6 × .6 × .6 × .6 = .08. So the most likely answer really becomes a very UNLIKELY occurrence. Hertz calls it an "unlikely coincidence." You might wish to call it the "unlikely most likely" case! The most likely case as a single answer is seldom what a decision maker really needs. The principal reason is that a single answer obscures the full spectrum of possible occurrences. Thus, the true, total range of possibilities is hidden. Most likely estimates, then, seldom tell the manager the extent of the risk he is taking.

Hertz lists five attempts made to improve forecasts or investment analysis and lists the flaws in each. He presents his own suggestions for "sharpening the picture" which are:

1. Estimate the range of values for each key variable.
2. Sample at random from the distributions of each variable. (With computers to help, from 500 to 5,000 iterations will usually produce an adequate final distribution reflecting the forecast or investment opportunity.)

Estimates for each variable are obtained from the most knowledgeable persons. In this book frequently reference is made to the triangular distribution because of its simplicity—both in using and in "extracting" information from the experts. As Hertz points out, guessing about a range is usually more accurate than guessing at a single number. A range, and particularly a three point range of minimum, most likely, and maximum, provide better basic data about the experts' beliefs involving a particular uncertainty. Such an estimate is far superior to an average. Even though an average represents the mean of your expectations, it does not fully describe the full suite of outcomes.

Hertz closes with "to understand uncertainty and risk, is to understand the key business problem and the key business opportunity." This classic article is recommended reading for students of risk analysis.

Hanssmann

A 1968 publication of John Wiley and Sons by Fred Hanssmann is entitled "Operations Research Techniques for Capital Investments." Of particular interest in this volume is Chapter 6 on "Techniques for Subject Estimation and Forecasting," page 211.[3]

Hanssmann makes the following points relative to subjective probability:

1. Subjective probability is denoted by the absence of a model that produces the desired answer from existing data.
2. Gaining data from experts (a key ingredient in any investment decision involving vast uncertainties) requires standardized and unambiguous questions.
3. Single estimates (most likely) should be used ONLY if you think uncertainty is either negligible or unimportant.
4. The most simple form of illustrating uncertainty is the distribution. It is a probabilistic approach which corrects the defects of other techniques that mislead the decision maker.

Reilly and Johri

A third reference by Park M. Reilly and Hari P. Johri[4] deals with the extraction of data from experts. The authors review the many real life situations where subjective probability exists. One example used is the physician and a single patient. For the patient no repeats of an illness are possible to help understand the odds for a cure. (An explorationist faces the same problem with only one prospect. No two oil or gas prospects are exactly alike and past experience is usually not precise enough to guide with absolute accuracy his decision to drill.)

In extracting odds, Reilly and Johri remind us that equal odds imply *no* ability to distinguish between alternates. Long odds imply considerable knowledge of the outcome.

Opinion analysis, say Reilly and Johri, can cover the entire range of outcomes plus indicating the degree of belief in the occurrence of each possible event. It is the combination of the two factors—range and probability—which help the decision maker.

Some Dissent

Not all managers approve of quantifying opinion. A few scientists even question its use—not because of the validity of the concept, but from errors in using the technique. A good example of these reservations was an article by Tversky and Kahneman[5] of the Psychology Department of Hebrew University, Jerusalem, Israel. Their major point is that heuristic (the self discovery process of learning; i.e., experience related learning) principles are useful but they can lead to either distortions from biases or to systematic errors. Their reservations include:

1. Lack of sensitivity of heuristics to prior probabilities of outcomes. Ignoring others' experience.
2. Insensitivity to sample size. Small samples can and do vary more from the mean of a large population. Remember our penny tossing experiment?
3. Misconceptions about chance. Ignoring the fact that a run of heads has no relationship to the next coin toss—it still remains a probability of 1/2 for heads.
4. Insensitivity to predictability and misconceptions about correlations.
5. Misunderstanding of the question asked or the goal sought.

Why Subjective Probability?

With these posed limitations one might well ask "Why subjective probability?" The answer to this question is—because there is nothing better. In spite of seeming limitations, opinion analysis is used and will continue to be used for this basic reason. Opinion analysis also has several plus factors working for it.

1. It can or should consider all critical factors bearing on the final judgment. At least all factors contemplated can be shown by documentation. It is of real value to know what *was* considered a key variable. "Seat of the pants" intuitive judgments leave no tracks! This does not mean that they are not often right—it is just that no one knows what went into making the decision.
2. In considering several related events, investments, risks, it helps to have the consistency which well documented opinion analysis can provide.
3. Up-side and down-side risks and relationships are better illustrated and handled. The complete spectrum of possibilities (beliefs) can be shown by documenting an analysis of opinion properly.
4. The degree of unknownness can be fully exposed—or expressed.

In a complex problem, intuitive judgments have a greater chance for error and inconsistency. Some methodology is needed to document fully the vital factors. It is for this reason that managers do turn to opinion analysis to aid them in making the right decisions under great uncertainty. The proper analysis of opinion is not an absolute guarantee of success; but in the long run it is a better way than unrecorded, inconsistent, intuitive judgment.

Two Examples

To demonstrate the usefulness of well documented analysis of opinion and what it can do to help in making decisions, two examples follow. Each deals with subjective probability in a simple straightforward manner. Both examples will involve prospect evaluation—the most perplexing problem facing the explorationist.

All prospect evaluations are a search for better understanding—a clearer meaning to a problem involving much uncertainty. Most explorationists would agree that the search is laudable even though they must use subjective probability (or some variation thereof) to arrive at their decision.

One example will deal with the chance of hydrocarbons existing; the other will deal with the possible sizes of a prospect if it is productive.

Chance of Success

First, let's deal with the chance of success. This part of risk analysis has many names. It is sometimes referred to as geologic risk; or geologic success; or existence risk; or the chance of adequacy (of hydrocarbon presence). Whatever the name, it is an attempt to assess the risk of hydrocarbon presence when the facts for assessment are *few*.

How do you go about estimating geologic risk? First, you must decide what factors control the existence of hydrocarbons. In Exploration Economics[6] three factors were used. They were:

1. Structure
2. Reservoir
3. Environment

The three factors ask these questions:

1. Are structural or stratigraphic conditions adequate for entrapment?
2. Are good reservoir conditions present?

3. Was the paleo-environment appropriate for hydrocarbon accumulation?

Few prospects allow three such simple questions. You may think of eight or ten or more factors influencing hydrocarbon presence. Fine. Remember, however, you must have data or experience for estimating each factor; otherwise, you are just grabbing numbers from the air. The way to solve the dilemma of too many factors for which no data are available is to combine factors for probability estimation. The term "environment" obviously includes many geologic variables important to hydrocarbon generation. However, it might be easier to think of these variables in their entirety than to make individual estimates. The three factors will suffice for the examples in this chapter.

How Many Factors?

The number of factors critical to hydrocarbon formation, entrapment and preservation can be considered very numerous. However, when the facts are few, we are better off to deal with as few variables as possible—otherwise, the tendency is to overwork the problem.

In addition there is a mathematical reason for limiting the number of variables. Since we are using the multiplication rule in probability[6] the greater the number of factors, the smaller will be our composite success estimate. In fact, the number of variables can be so large that the success estimate can become unrealistic. This result occurs even though the risking is not severe. Consider the following two examples where each factor is risked at .9 or .5—i.e., 90% or 50% chance of adequacy.

In the multiplication rule of probability—the probability of two or more independent events having specific outcomes is the product of their separate probabilities. In our simple example of Table 9.1, we have assumed that we are dealing with independent variables and that each variable has the same chance of occurrence—i.e., either 90% or 50%. As the number of factors increases the composite geologic success estimate decreases, even at such low levels of risk for each factor as 90%. Too many factors, then, can produce an answer well below empirical experience. Ten variables or factors each risked at 50% produce a final product of only .1%—a probably unrealistic number. Four factors each risked at 70%, not shown on Table 9.1, yield a geologic success estimate of 0.24.

A reality check will thus be necessary for setting up the basis for a geologic success estimate. If you are in a trend where the average

TABLE 9.1

THE EFFECT ON GEOLOGIC SUCCESS ESTIMATE OF AN
INCREASING NUMBER OF FACTORS

Number of Factors	Composite GSE	
	@ .9	@ .5
3	$(.9)^3 = .729$	$(.5)^3 = .1250$
4	$(.9)^4 = .656$	$(.5)^4 = .0625$
5	$(.9)^5 = .590$	$(.5)^5 = .0313$
6	$(.9)^6 = .531$	$(.5)^6 = .0156$
7	$(.9)^7 = .478$	$(.5)^7 = .0078$
8	$(.9)^8 = .430$	$(.5)^8 = .0039$
9	$(.9)^9 = .387$	$(.5)^9 = .0020$
10	$(.9)^{10} = .349$	$(.5)^{10} = .0010$

success rate is 30% then your composite geologic success estimate (GSE) should yield about 30% for a series of prospects. If the series will not average this amount you are either dealing with a set of prospects different from the discoveries to date (a possibility); or, you are out of touch with reality. You must decide.

If the trend average is 30%, why risk each prospect? The reason is very simple. Always remember that a success rate is an average. An individual prospect is one event. It cannot be an average. You should risk each prospect so that a series can be developed for an average. Geologic judgments should enable you to sharpen your estimates about a particular prospect. You must decide if this prospect or that prospect has the best chance of being productive.

Not to believe that you can select, on the average, the best prospects for drilling is to agree that the trend average will automatically apply to your prospect, or that your geologic judgments cannot isolate the better than average prospects. Either approach can lead to a high cost of finding and a large possibility of going broke. Risking individual prospects, whether intuitively or formally, is an important part of managing an exploration program.

Gehman, Baker & White[7] have proposed a technique, similar to that just described, in a paper given at an AAPG Research Symposium. This paper is recommended reading for all who are interested in prospect risk analysis.

The Results of More than One Estimate

The data on critical geologic parameters in the search for hydrocarbons, illustrated in Exploration Economics, were based on an assumed sample of 50 exploratory wells.[6] These wells present an empirical base for future reference in estimating geologic success in the same play for the remaining undrilled prospects.

Using the three parameters from Exploration Economics, one person can calculate a geologic success estimate for a prospect. You then have a single opinion. Suppose, however, that you want more than one opinion. Such a desire would be expected if the money involved in a decision were large. How would you gather, assemble and handle several estimates and what would they indicate?

On Table 9.2 the results of 10 separate estimates are shown for the three parameters with the composite geologic success estimate calculated for each.

TABLE 9.2

GEOLOGIC RISK FACTORS
PROSPECT A

Name of Geologist	Risk Factors[a]			Composite GSE[b]
	Structure	Reservoir	Environment	
Able	.5	.3	.9	.135
Baker	.7	.5	.8	.280
Charlie	.4	.7	.7	.196
David	.6	.2	.9	.108
Edgar	.8	.9	.8	.576
Frank	.7	.5	.7	.245
George	.6	.5	.6	.180
Hector	.7	.4	.7	.196
Ignatio	.8	.7	.6	.336
Julius	.7	.7	.7	.343
Average	.65	.54	.74	.260

[a] Estimated chance of favorable occurrence.
[b] GSE = geological success estimate.

For example, geologist Able estimates as follows:

	%	Prob.
The chance of the structure being as he sees it	50	(.5)
The chance of getting good reservoirs	30	(.3)
The possibility of an optimum depositional environment	90	(.9)

Able's composite geologic success estimate (GSE) is the product of the three "independent" parameters.

$$(.5) \times (.3) \times (.9) = .135$$

He estimates only a 13.5% chance for success for Prospect A.

Able's estimate is not the lowest, however. That honor goes to David with a GSE of .108. David's estimate is low because he places a high risk (only 20% chance of adequacy) on getting good reservoir conditions.

The estimates of Able and David illustrate an important point—the smallest estimate of risk *controls* the answer! You can easily understand this point by considering what happens if the estimate of success for any of the three parameters is zero. Then the GSE is also zero—because if you think there is no chance for good reservoir conditions then you have no prospect and the GSE should be zero. So, the lowest estimate of adequacy controls the GSE.

Edgar wins the prize as the optimist; he has high values for each parameter:

.8 for structure
.9 for good reservoirs
.8 for optimum environment

Edgar's probabilities, when multiplied, yield an answer of only .576. Even relatively high estimates for the parameters will produce a lower composite because you are saying *each* must happen favorably for the final outcome to be successful; therefore, the risks compound whenever you have two or more events in which all must occur for your desired outcome. Remember rolling two sixes with a pair of dice has a much lower probability than rolling one six from two dice.

If you want to summarize the ten estimates you can add the 10 GSE's and divide by ten. The average of all ten GSE's is .260; you can get the same result by summing the ten estimates under each parameter, averaging each, and multiplying the averages:

$$(.65) \times (.54) \times (.74) = .26$$

The consensus of our ten geologists is a 26% chance—about one in four—of the prospect being productive.

Our Favorite Distribution

A closer examination of the data reveals some other interesting facts. Seven of the ten values are .28 or lower—many low estimates and few large ones. That statement immediately makes us suspect a lognormal distribution! Sure enough, we have a rough approximation of a lognormal distribution from our data. See Fig. 9.2. Once again we have shown

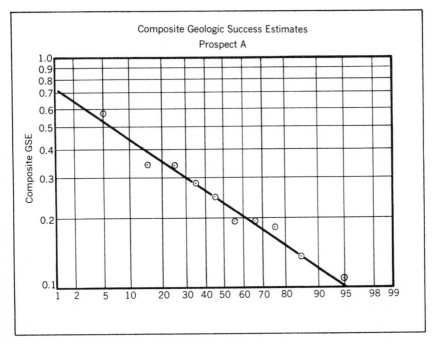

FIG. 9.2

that a series of products from several factors will yield a lognormal distribution.

To refresh your memory on the construction of a distribution plot on lognormal probability paper, the plotting points are shown in Table 9.3.

TABLE 9.3

GSE PLOTTING POINTS FOR FIGURE 9.1

GSE Value	Cumulative Probability Midpoints—%
.576	5
.343	15
.336	25
.280	35
.245	45
.196	55
.196	65
.180	75
.135	85
.108	95

Other Observations

Six values are less than our mean of .26 and four are larger, again demonstrating that the mean of a lognormal distribution is skewed (toward larger values) from the median which on our figure is .23.

The sum of the estimates for the ten geologists shows the highest value for environment—.74. There is the least divergence of agreement for this parameter. Wide divergence produces the lowest average for estimates of reservoir condition. For each parameter the average and range is shown below.

	Average of Ten	Range of Values
Structure	.65	.5-.8
Reservoir	.54	.2-.9
Environment	.74	.6-.9

Again, it is the low values, not the range per se, which produce the low average.

The Meaning

Suppose you decided to use the consensus of several explorationists to arrive at your final GSE. What should you expect and what is its meaning? Is any further action required?

The biggest value of considering each parameter separately is that you can see where the differences are; i.e., you can see where your estimators agree that the uncertainty is the greatest.

The first check you would make is a re-examination of the factor on reservoir conditions. Obviously some of the estimators have concerns here. The range of opinion is the most broad and the average chance of favorable conditions is the lowest. The consensus, then, is that reservoir conditions are the most critical—the most uncertain. Additional work may need to be done in this area including a review of all data for possible further definition.

You would also want to make a reality check for the consensus against the average success rate of the trend or play. Although the same number would be coincidental, this reality check should alert you to consider further if the result is widely divergent from the trend average; that is, are your prospects *really* that different?

Field Size

Our first example dealt with a simple method for estimating *if* hydrocarbons are present. This question is the first which must be answered by the explorationist.

His next logical question, after a favorable answer on hydrocarbon presence, would involve size. He would ask "How big is it?"

This question is of vital importance. Why? It is vital because of the unusual distribution which assures many small fields and few large ones. The next phase of risk analysis should deal with methods for determining or estimating size. After all, you want to look only for the larger fields, if possible, so a documentation of your concept of size is most important.

The example will be drawn from Exploration Economics[6] and will show, again in simple fashion, how to document a range of opinion to show the full spectrum of reserve sizes possible for a prospect.

Reserve Estimates and the Three Factors

Every oil or gas field has a certain areal size, a thickness of the producing formation and a recovery per foot of pay. For a single-reservoir field the reserve could be calculated by:

Net pay (ft) × recovery per acre-ft (bbl) × area (acres).

The result is a reserve estimate in barrels, often expressed as a single number. Subjective probability is used in prospect evaluation because we do not have absolute answers for the three factors of pay, recovery and area.

Under such circumstances consider what values appear reasonably possible. In the example from Exploration Economics a three-value suite was estimated for each parameter and a probability of that value occurring assigned. Table 29 in Exploration Economics shows these values and their probabilities.

Parameter	Value	Probability
Net pay—ft	50	.3
	100	.5
	190	.2
Rec/ac-ft—m̄cf	0.7	.3
	1.0	.6
	1.5	.1
Prod. area—acres	800	.3
	1,000	.5
	1,400	.2

The various combinations of these parameters produce 27 separate estimates, each of which has its own specific probability of occurrence.

Plotted on lognormal probability paper, Fig. 9.3, the 27 cases show the relative chance of occurrence of fields of a given size or greater.

Subjective probability was used here, not on the final answer, but on the key ingredients producing the answer. The opinions about the possible range of values for each parameter were taken into account. The result is a full suite of values showing the complete spectrum of possible cases—plus their possibility of occurrence. These two facets give increased dimensions of the opportunity to the decision maker. They are far superior to a single answer—which for a given prospect has almost zero chance of occurring.

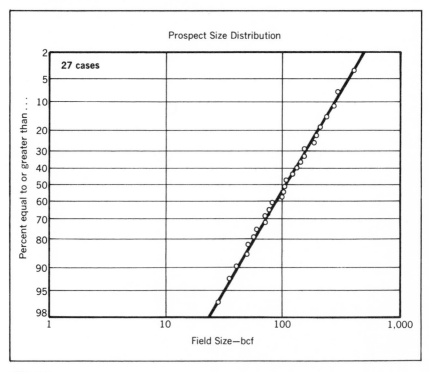

FIG. 9.3

The brief review of a use of subjective probability from Exploration Economics illustrates another way to document the opinion of experts. We described a needed quantity where a single answer would not disclose the range of possible answers, much less their possibility of occurrence.

In this last example, the three values for each parameter could be converted to a triangular distribution. Triangular distributions produce the opinion of the experts *without* asking them for a guesstimate of probability. Just how this can be accomplished is the subject of the next chapter on triangular distributions.

REVIEW

Whenever an event has many possible outcomes, a single answer hides much from the decision maker. The analysis of expert opinion is a method which helps the decision maker by showing him both the full range of possibilities and their probability of occurrence. These two products of subjective probability enable better and more consistent decisions. Furthermore, the documentation possible will allow an after-the-fact analysis of a series of decisions to aid future choices. Intuitive judgments are often right, but they have a much greater chance for error; they can be inconsistent relative to complex problems, and they "leave no tracks."

BIBLIOGRAPHY

1. Hertz, David B., "Risk Analysis in Capital Investment," Harvard Business Review, Jan.-Feb., 1964, pg. 95.
2. Hertz, David B., "New Power for Management," McGraw-Hill, New York, N.Y., 1969.
3. Hanssmann, Fred, "Operations Research Techniques for Capital Investments," John Wiley & Sons, New York, N.Y., 1968.
4. Reilly, Park M. and Johri, Hari P., "Decision Making Through Opinion Analysis," Chemical Engineering, April 7, 1969, pg. 122.
5. Tversky, Amos and Kahneman, Daniel, "Judgment Under Uncertainty: Heuristics and Biases," Science, Vol. 185, Sept. 17, 1974, pg. 1124.
6. Megill, Robert E., "An Introduction to Exploration Economics," Petroleum Publishing Co., Tulsa, Oklahoma, 1971, pg. 98-106.
7. Gehman, H. M., Baker, R. A., White, D. A., "Prospect Risk Analysis," *in* Davis, J. C., Doveton, J. H., and Harbaugh, J. W., Convenors, Probability Methods in Oil Exploration—Amer. Assoc. of Petroleum Geologists, Research Symposium, August 20-22, Stanford University Preliminary Report, 1975, pp. 16-20.

10 What You Need to Know About Triangular Distributions

Assigning probabilities to the possible range of values of pay, recovery and area may make some experts uncomfortable. Yet you still need to quantify the experts' opinions in a usable form. The triangular distribution often represents the answer to this problem. It automatically assigns probabilities, as will be shown in the following chapter, yet the word "probability" need not be mentioned.

The Input

All you need for a triangular distribution is three values:

—a minimum
—a most likely
—a maximum

These three values must be carefully defined as they are often misunderstood.

For triangular distributions the minimum is the absolute number below which no value can exist. The frequency for the minimum is zero! If you ask a geologist what the minimum thickness is expected to be, make sure he *knows* he is saying no value can exist below the minimum. The reason this definition is *stressed* is simple. Many experts will give a value for a minimum and will mentally think that that value could occur, but rarely (i.e., has a low probability of occurrence). However, in a triangular distribution the minimum means *zero* probability. So the minimum will *never* occur. If the expert thinks his minimum can occur, *choose a lower minimum.*

The most likely is that value which should occur most frequently. It is the mode of the distribution—not the mean. Since we deal with lognormal relationships, the most likely value will be nearer the minimum than the maximum.

The maximum value, as in the case of the minimum, is an absolute ceiling. No value larger than the maximum can exist. Likewise, the probability of the maximum value itself is zero. So, if an expert thinks of his maximum as some value which could occur—set a higher maximum. In other words, make sure that he knows the maximum can *never* occur.

A Triangular Distribution

Let's suppose that you are asking an explorationist to estimate the possible productive area for a prospect. You ask him to pick an "absolute" minimum, an absolute maximum and a most likely size. His answers might be 500, 2000, and 1000 acres. He would most probably have geologic reasons for his choices or empirical ones based on experience. For example he might say:

1. One structure in the trend was filled to the spill point (the last closing contour). On this prospect that would mean 2,000 acres.
2. On the other hand, a few are only 20 to 25% filled. If the prospect is like that, it will be only 500 acres.
3. However, I really think the field will be contained in 1,000 acres. That's the most likely estimate of size.

On Fig. 10.1 the triangular distribution for his opinions is drawn. For reasons that will be clear in the next chapter, no values are assigned to "*f*", the frequency associated with estimates. They are automatically determined when the value for the minimum, maximum and most likely sizes are set.

Note the ease with which the information was extracted from the explorationist. No probabilities were mentioned. Just his opinions were requested in relation to his experience. The same process can be used to estimate the range of values for recovery per acre-foot, and net reservoir thickness. The ease of extraction makes the triangular distribution one of the most popular of all opinion-defining distributions.

Assigning Probabilities

Suppose the expert does wish to assign probabilities? Then you have other factors to consider. Many pitfalls can occur in assigning probabilities, if you are not careful in the selection of values or probabilities.

To begin with, since the minimum and maximum in a triangular distribution have, by definition, values of *zero*, you cannot use these

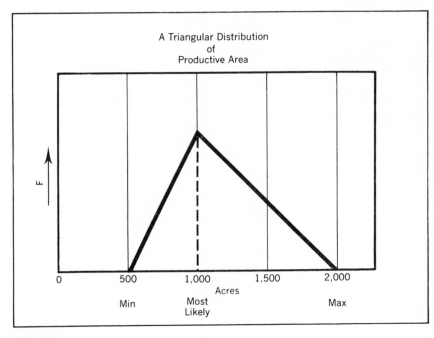

A Triangular Distribution
of
Productive Area

FIG. 10.1

terms. New terms, such as low-side, high-side, or low value and high value must be used to distinguish between those terms assigned a given probability and the maximum–minimum values.

One commercial evaluation program, POGO (an acronym for Profitability of Oil and Gas Opportunities),[3] uses the terms *minimum expected* value and *maximum expected* value. The point to remember is simple—if you assign probabilities to the low value or high value in a three point triangular distribution, avoid the terms minimum and maximum. In true triangular distributions, they must have the value of zero probability.

Nevertheless, you will not always see this terminology used and, therefore, despite the pitfalls, some discussion of assigning probabilities to the lowest and highest expected values is warranted.

What if a geologist or geophysicist thinks there is a 20% chance of the area being as large as 2,000 acres and a 20% chance of the field being as small as 500 acres? As you will see in Fig. 10.2, we now have a totally different distribution. Why?

First of all, in assigning probabilities to the low and high values, we have assigned a definite probability to the most likely; i.e. 100% − 20% − 20% = 60%.

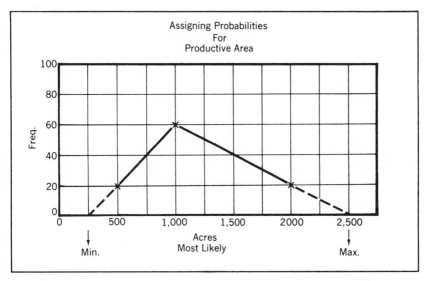

FIG. 10.2

Secondly, only by extending the 20% points to the horizontal base line are the true values of the maximum and minimum known. The maximum is now 2,500 acres and the minimum is 250 acres.

As we shall see later, Fig. 10.2 still does not describe the right probability but most programs for Monte Carlo simulation correct for this type of input.

Triangular distributions set up in the manner of Fig. 10.2 are sometimes called weighted distributions. The points to remember are:

1. Assigning probabilities for lower and upper range values does *not* give the same frequency distribution as saying the lower and upper values represent the minimum and maximum.
2. Having assigned two probabilities in a triangular frequency distribution, we have automatically determined the third probability since the three must add to 100.

You can combine the two techniques and assume a value for either the minimum expected or the maximum expected value but *not* both. You would get a triangular frequency distribution such as shown in Fig. 10.3 if you assigned a 20% probability to the chance of 2,000 acres.

Note that since *zero* probability is assigned to our minimum value and 20% to our maximum value, the most likely value becomes 80%.

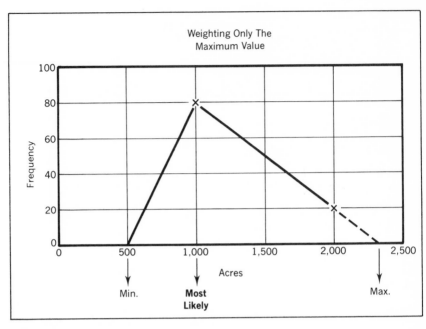

FIG. 10.3

Projecting the line through the 20% chance for 2,000 acres, we arrive at a maximum value of 2,325 acres.

Now compare Figs. 10.2 and 10.3. It may surprise you to know that the mean of the distribution in Fig. 10.3 is greater than the mean of Fig. 10.2. Even though the maximum value in Fig. 10.3 is less than the 2,500 of Fig. 10.2 the mean value is 1,274 for Fig. 10.3 compared to 1,251 for Fig. 10.2.

Normally increasing the maximum value will increase the mean; but in Fig. 10.2 we *lowered* the minimum value also.

Changing the Maximum Value

We will have more to say about mean values in the next chapter; first, however, let's show for our base frequency diagram the effect of changing the maximum values.

On Fig. 10.4 our base distribution is shown as maximum value ①. For reasons to be shown later it is marked "lognormal". The mean value of this triangular distribution is 1,170 acres. A maximum value

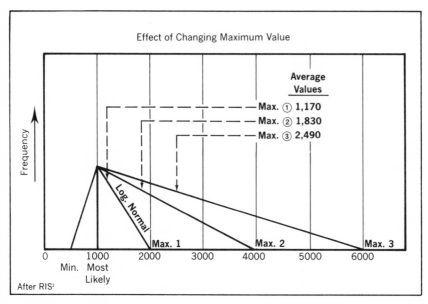

FIG. 10.4

of 4,000 acres (max. ②) increases the mean to 1,830 and a maximum value of 6,000 (max. ③) increases the mean to 2,490 acres.

Thus, increasing the maximum value (leaving the minimum and most likely values the same) increases the mean value.

Changing the Most Likely Value

Changes in the most likely value, leaving the minimum and maximum values the same, also changes the mean—in the direction of change of the most likely value. On Fig. 10.5 the base distribution is shown with ① at the base case most likely value of 1,000 acres. Lowering the most likely value ② lowers the mean. Raising the most likely value, ③ and ④, increases the mean value.

Triangular Distributions and Histograms

Much earlier, in Chapter One, we reviewed the basic bar type frequency distribution called a histogram. You can also describe the

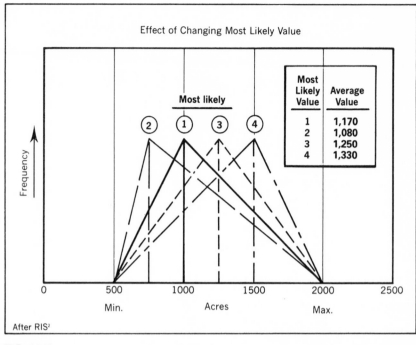

FIG. 10.5

variables in histograms, with the same data extracted from the expert for the triangular distribution. However, you must carefully observe one of the rules of constructing bar distributions or histograms—namely to keep the class intervals equal or constant.

Fig. 10.6 illustrates the area of a prospect in the form of a histogram. The class interval is 500 acres. We begin by assigning the midpoints of the first two intervals probabilities of 20% and 60%. However, we cannot arrive at a probability for 2,000 acres and a probability of 20% until we account for the class interval with the 1,500 midpoint which as drawn would have a probability of 40%.

By now, you have noticed that something is amiss. The probabilities add up to more than 1.00! Their sum is 1.40—an impossibility since three point probabilities (larger than the minimum and maximum for the extremes) must sum to 1.00. We solve this dilemma easily by dividing each probability by 1.4 (or multiplying it by the ratio of $\left(\dfrac{1.00}{1.40}\right)$. We now have a new probability for each class interval. These are shown in parentheses above each class interval and they sum to 1.00. This

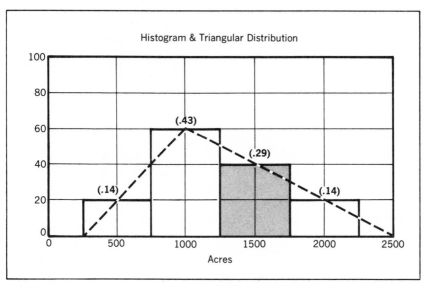

FIG. 10.6

process is much more cumbersome than the minimum, maximum, and most likely estimate of a triangular distribution; but it achieves about the same answer.

Actually, we had this same problem in Figs. 10.2 and 10.3 but most computer simulation programs which accept three points and their probabilities will automatically correct to the proper histogram.

One other pitfall possible with assigning probabilities for a triangular distribution is to choose a probability for the minimum expected value that results in the projection of the true minimum (zero probability) as being *negative*.[1] You can avoid negative (unreal) values for your minimum by specifying the minimum only with no probability assigned. That value becomes zero and cannot be negative.

If three values are adequate for gathering an expert's opinion, the triangular distribution with *NO* probabilities assigned is recommended. You can then avoid the problems which can accompany assigned probabilities. The triangular distribution is simple, quick and is easily converted by computer programs into a cumulative probability distribution—to be thoroughly reviewed in the next chapter.

Approximation of Lognormality

A good "rule of thumb" ratio will help you achieve or check for lognormality. In a normal distribution the distance between the minimum and most likely value is equal to the distance between the maximum and most likely. For a lognormal distribution the *log* of the two intervals is equal. So you can get an approximately lognormal relationship for your triangular distribution by satisfying this ratio for the three values:

$$\frac{\text{minimum}}{\text{most likely}} = \frac{\text{most likely}}{\text{maximum}}$$

You can also see from the relationship that the $(M.L.)^2$ is equal to the minimum value $\times$ the maximum value.

This relationship is not intended in usage to encourage you to "force fit" all of your triangular distributions to this ratio. Rather the value lies in its use as a reference point. If your parameter tends to be lognormal then you can check your three value triangle to see if you are thinking lognormally. It's a simple, useful ratio relationship.

REVIEW

One of the characteristics of the triangular distribution is that the values entered for maximum and minimum are *absolute* values.[2] The maximum cannot be exceeded and nothing can be less than the minimum. Furthermore, statistically, neither will ever be reached and values near the minimum and maximum will be rare. Many explorationists falsely assume that the triangular distribution allows for some small probability of occurrence of the minimum and maximum values.

As a result of reading this chapter, you should know the pitfalls in using triangular distributions as input for computer simulation models; particularly those where probabilities are assigned the low and high values. You can now avoid them.

The triangular distribution today ranks as one of the simplest and most effective ways of extracting data from an expert—if the number of possible answers is broad and the best guesstimate is underlain by much uncertainty. The triangular distribution is one of the tools that helps show the full spectrum of possibilities.

BIBLIOGRAPHY

1. Megill, Robert E., "An Introduction to Exploration Economics," Petroleum Publishing Co., Tulsa, Okla., 1971, pgs. 151-52.
2. Swanson, R. I., "More on Triangular Distributions," unpublished memo, October, 1972.
3. POGO is a computer program authored by PSI Energy Software.

11 Converting Triangular Distributions to Cumulative Frequencies

Triangular distributions are increasingly used as input in Monte Carlo simulation programs to express estimates of the limiting and most likely values of a variable. Such programs convert triangular input into cumulative frequency distributions.[1] From these cumulative frequency curves we can read probabilities for a specific value of the variable—yet the explorationist did not consider probabilities as part of his input! How is this possible?

Let's restate this question another way. You input into a simulation program three values for a variable—say pay thickness. You say: (See Fig. 11.2)

a minimum value is	10 ft
the most likely value is	50 ft
the maximum value is	250 ft

The simulation program converts these three values into a cumulative frequency curve which expresses probabilities of occurrence. The nearest word to probability used by the explorationist was most likely! How then does most likely become a specific probability? What process is used in the computer program to convert three simply stated values into specific probabilities?

The Logic of Conversion

The manner in which this conversion is accomplished first starts with the entire triangle and equates the area of the triangle to 1.0. You must now recall another fact. The sum of all the probabilities of occurrence of an event must also equal 1.0. If this is true, the sum of all the probabilities of net pay, in Fig. 11.1, must equal one (1.0). Each value of net pay will have some frequency of occurrence, such as points P and f—where:

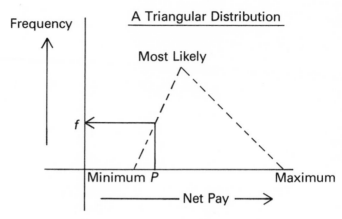

FIG. 11.1

> P represents any pay thickness and
> f represents its corresponding frequency of
>   occurrence.

We started our discussion by stating that we assign values only to P but derive f on a cumulative basis!

Any histogram can be converted into a cumulative frequency distribution. We did this in Chapter One. You simply convert each class interval to percentage of the total and then add the class interval percentages of the histogram cumulatively. A triangular distribution is a form of histogram. The area under the triangle represents the frequency of all of the events which can occur. Greater frequencies have greater area. Just imagine a large number of class intervals each drawn to the upper line of the triangle and you can visualize how larger frequencies have larger areas. One can use area then, to express frequency and that is the key to the mathematical conversion.

The Math of Conversion

If you are not interested in the mathematical concepts behind the logic of conversion, you may wish to move immediately to the section in this chapter headed "Practical Applications."

To convert P into cumulative values of f requires the introduction of another concept. A *small* triangle is formed by drawing a line perpendicular to the base of the larger triangle. The area of the small triangle represents some proportion of the area of the entire triangle.

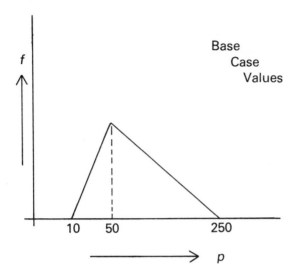

FIG. 11.2

By prior logic it then also represents some value less than 1.0. Since *P* values with larger frequencies will have larger triangles, the area of the triangles can be used to simulate cumulative probabilities.

To do this and involve only base line values requires some awkward, but simple, math; and it so happens that the ratio of any *portion* of the total area to the total area can be expressed as a function of values of the base line. This means you can develop a cumulative frequency curve for the distribution of possible pay thicknesses using only *three* values of pay thickness; namely,

> the minimum
> the most likely, and
> the maximum.

The actual mathematical proof for this premise is detailed in Appendix A. We shall use the end result to show what it can teach us about the three values of minimum, maximum, and most likely.

The Equations

Please begin this section by referring to Fig. 11.3 which is the reference triangle for the first equation.

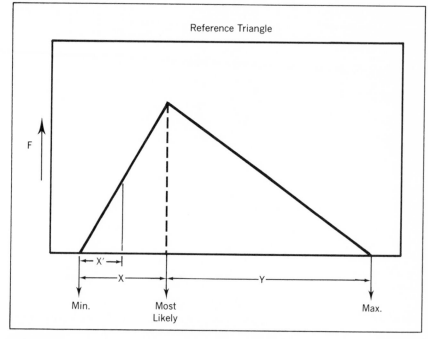

FIG. 11.3

Number One—To the Left of the Most Likely

Two equations are needed to develop our cumulative frequency curve. The first deals with those values of net pay to the left of (smaller than) the most likely value. The equation is:

$$\underset{x'=0}{\overset{x'=x}{C F}} = \frac{\dfrac{(x')^2}{x}}{(x+y)} \qquad\qquad 11.1$$

where, CF   is cumulative frequency

x' is any value of net pay along the base line from the minimum to the maximum value. Numerically, x' starts from *zero* at the extreme left of the triangle, i.e., from the minimum value.

For example: In the specific case (Fig. 11.2) at the most likely value of 50, x' has a value of 40—i.e., 50 minus 10.

x = the value between the minimum and most likely value
(the most likely value minus the minimum).

y = the value between the most likely and the maximum.
(the maximum minus the most likely value).

Putting the mathematics of this equation into words we would say—for a triangular distribution, the cumulative frequency from $x' = 0$ to $x' = x$ is equal to x' squared, divided by x and that quantity divided by $(x + y)$.

Note what happens when $x' = x$. Then the numerator reduces to x and the equation becomes:

$$\underset{x'=0}{\overset{x'=x}{C\,F}} = \frac{x}{x + y} \qquad\qquad 11.2$$

In Fig. 11.2 (for values 10, 50, and 250),

$$x = 40 \quad y = 200 \quad \text{and therefore} \quad x + y = 240.$$

Using the values from this example, note that the value of x (most likely minus the minimum) divided by $(x + y)$ is:

$$\frac{40}{240} = \frac{1}{6} = .167$$

The number 0.167 is valid only for our special case of input (10, 50, 250).

This ratio tells us that 16.7% of the values are equal to or less than the most likely value and 83.3% are larger. The significance of this statement is—even though the value of 50 is our most likely value, the range of values is so great and the distribution skewed so far to the right that 83.3% of the values of net pay are *greater than* the most likely—greater than the one value said to occur with the greatest frequency.

You can illustrate this point by drawing small rectangles along the triangle (keeping the class intervals equal); you will see visually that most cases lie to the right of the most likely value. You can also prove it by considering only class intervals of value 10. There are four such intervals to the left of the most likely and 20 to the right. There are five times as many class intervals to the right as to the left!

The first significant point is this: when the most likely value is much nearer the minimum than the maximum (when the distribution is skewed to the right) the most likely value will have a small cumulative frequency. Mathematically we are saying (where $x' = x$) if x is small relative to y, the ratio $\dfrac{x}{x + y}$ will also be small. (The value of y relative to x is a measure of skewness. Values of y much larger than x mean greater skewness).

This significant point sounds strange doesn't it—the most frequent value being in the smaller end of the cumulative frequency distributions. The cause is, of course, the wide range from the minimum to the maximum. The broader this range, the smaller likelihood of occurrence of any single value, including the most likely value. Even though the most likely value will occur more often than the other values, a broad range makes all individual frequencies small.

We can show this another way by looking at an isosceles triangle. Under these conditions $x = y$ and so $\dfrac{x}{x + y}$ = .50. Thus our most likely value occurs at a cumulative frequency of 50%; i.e., half the values are smaller and half are larger!

Perhaps these illustrations and examples will help you begin to understand some of the interesting aspects about triangular distributions. Remember, however, that lognormal distributions—those most commonly found in field size distributions and other reserve related parameters—are *always* skewed to the right; that is, the most likely value (the mode) is always nearer the minimum than the maximum.

Number Two—To the Right of the Most Likely

The next equation is more complex. For values of the parameter to the right of the most likely (net pay in our case) the cumulative frequency can be calculated by the following equation:

$$\mathop{\mathrm{CF}}_{\substack{x'=(x+y) \\ x'=x}} = 1 - \frac{\left(1 - \dfrac{x'}{(x + y)}\right)^2}{1 - \dfrac{x}{x + y}} \qquad 11.3$$

This equation looks very formidable. However, in use it is not. The denominator of the fraction $\left(1 - \dfrac{x}{x + y}\right)$ is a constant for any three values, so it need be calculated only once.

For the base illustration of net pay, Fig. 11.2, the denominator becomes:

$$1 - \frac{40}{240} = 1 - .167 = .833$$

This equation (11.3) represents the area to the right of the most likely value *out to x'* plus that to the left (i.e., areas $a + b$) as shown on Fig. 11.4. Note the entire fraction subtracted from 1. It represents any small triangle (DEF) created by a value of x' less than $(x + y)$ (area c). As x' approaches $(x + y)$, the area of the small triangle approaches zero. When $x' = x + y$, the area = zero.

What happens when $x' = x$? In words, when $x' = x$ our expression (ratio) represents the entire triangular area to the *right* of the most likely value or areas $b + c$ in Fig. 11.4. Again, looking at the ratio only (the part subtracted from 1) and substituting x for x' our ratio reduces to $1 - \frac{x}{x + y}$. From our first equation we remember that $\frac{x}{x + y}$ is the area of the entire triangle to the left of the most likely value (area a). We now know that $1 - \frac{x}{x + y}$ is the area of the entire triangle to the right of the most likely value $(b + c)$. With just a little intuition, you may have guessed that the area to the right of the most likely value is also equal to $\frac{y}{x + y}$.

If that is so, then $1 - \frac{x}{x + y} = \frac{y}{x + y}$.

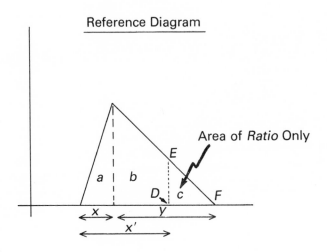

Reference Diagram

FIG. 11.4

Rearranging we have $\dfrac{x}{x+y} + \dfrac{y}{x+y} = 1.0$.

In words, this statement says: the area to the left of the most likely value plus the area to the right is equal to the area of the whole triangle. By definition, at the outset, we set these areas equal to one.

One can also see from this relationship that the bottom part of the ratio $\left(1 - \dfrac{x}{x+y}\right)$ in the second equation could be replaced by $\dfrac{y}{x+y}$.

Practical Applications

I. Changing the Input.   Up to this point we have demonstrated that areas (cumulative frequencies) of the triangle can be expressed as values of the base line alone. Now let's put this information to some practical use.

Fig. 11.5 is the first of several illustrations from which we will learn

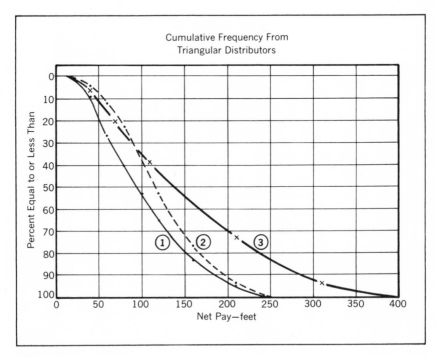

Cumulative Frequency From
Triangular Distributors

FIG. 11.5

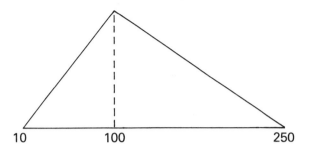

FIG. 11.6

more about triangular distributions. Three curves appear on the figure. They are adapted from ideas of E. T. Lewis[3] covering this particular point.

Curve 1 represents the cumulative frequency distribution for the base case—Fig. 11.2.

Curve 2 represents the cumulative frequency distribution of the same range of data but a *different* most likely value. This distribution is:

From a comparison of curves 1 and 2, three observations can be made:

1. A doubling of the most likely value, from 50 to 100, did not produce a drastic change in shape of the cumulative curves.
2. Since the minimum and maximum values did not change, the major effect is in the lower and middle values of the cumulative frequency. In a right skewed distribution one would expect the change here, since we are changing the locale of the most likely values.
3. At the 50th percentile (the median—not the most likely) the most probable thickness changed from 95 ft to 115 ft, an increase of 21%. At the 90th percentile there is only a difference of 10 ft, an increase of only 5%.

Curve 3 is a third triangular distribution in which the maximum value has been increased. This distribution is (Fig. 11.7):

The maximum value is now 400; the minimum and most likely values stay the same. Some observations here are:

1. We see no steep slope on Curve 3 which means no dominant frequency. Remember, steep slope on a cumulative curve means high frequency.

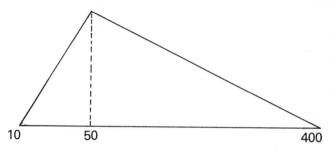

FIG. 11.7

2. We affirm point one by observing our triangle. The wide range of values (class intervals) allows no single value or class interval to occur at a very high frequency.
3. Percentage-wise the change in the most likely value was greater than the change in the maximum value. Doubling the most likely value changed the 50th percentile to 21%. We did not even double the maximum value but it changed the 50th percentile by *47%*. At the 90th percentile, the change was significantly greater. We would expect this trend since we changed only the maximum value.

In summary, changing our input demonstrates that changes in the maximum or minimum value produce greater variance in the cumulative frequency curve than an equivalent change in the most likely value. This is an important point to remember if you use triangular distributions as input for a computer program using Monte Carlo simulation. The extremes (maximum and minimum) exercise the greatest control over the cumulative frequency curve.

The changes outlined have the same effect on the mean (average) value for the curves. The means are as follows:

Curve	Mean (Feet)	Change from Curve 1 – %
1	99	—
2	114	15
3	147	48

II. The Locked-In Frequency. Now that we have shown that frequency can be calculated from base line values alone, we can review both verbally and visually another aspect about triangular distributions. Setting the three values, minimum, maximum, and most likely, *locks in the frequency.* Visually this observation produces the surprising conclusion that the height of the triangle has no bearing on the cumulative frequency distribution.

Consider the following triangle (Fig. 11.8).

We have just said that the cumulative frequency distribution for triangle ABC is identical to that for *ABD*! Surprises you, perhaps; but it's true. Remember we have proved mathematically that the areas (which represent frequencies) resulting from any value along the base line (between minimum and maximum values) can be calculated from values along the base line alone.

How is this possible? Mathematically we show, in Appendix A, that all values of h (height) drop out of the equation. Visually, however, we can see that each triangle produced by a value of x' will represent the same identical proportion of the total area of the triangle of which it is a part. Granted, triangle ABC has a larger area than triangle ABD; but it also has larger proportionate triangles (areas) produced by the same values of x'—so that *the ratio remains constant*. One other important fact comes from this analysis. You set the frequency of the most likely

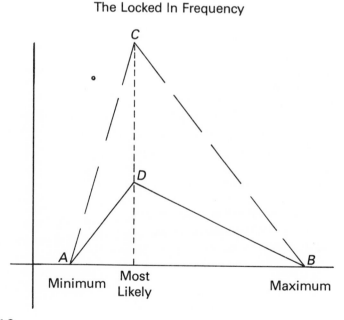

The Locked In Frequency

FIG. 11.8

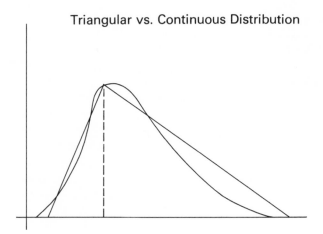

Triangular vs. Continuous Distribution

FIG. 11.9

value when you set the extremes—the minimum and maximum values. Once set you have a locked-in maximum frequency. You can change the distribution shape by moving the locale of the most likely; but its maximum frequency remains the same.

Thus, we have another important point.

III. Triangles Are Not the Real World. We must not forget one important point before we leave this subject. Triangles do not give precise representations of either normal or lognormal distributions. For example, note the following comparison (Fig. 11.9).

In this comparison the triangle misrepresents both sides but particularly the right side of the real world distribution. Why do we use them? Because it's the only way to approximate a distribution from only three points—minimum, maximum, and most likely. In areas of great uncertainty, estimating even three values cannot be done with accuracy. So we still use triangular distributions and they are accurate enough as computer input for uncertain parameters.

Because much of the input should be lognormal, the greater distortion exists on the high side. Moreover, triangular distributions tend to overestimate the mean value compared to a real world lognormal distribution.

For most of the applications we have reviewed these distortions are not serious. They become so if y becomes very large relative to

x. Thus, if any of the input triangles show large skewness to the right (large values of y) keep this distortion in mind.

What Does All This Mean to the Explorationist?

Triangular distributions are very useful mechanisms for input into Monte Carlo simulation. They allow an approximation of a real distribution from only three points. They do, however, tend to distort the real world when the data are heavily skewed to the right. When the potential range of values can be great, the distortion will be greatest.

Such a fact may be disturbing. Remember, however, that a wide range of possible events also expresses our own uncertainty as to what might occur. Thus, the distortion may be less than the accuracy of the input. It is not serious for lognormal relationships.

REVIEW

We use many tools which have limitations. As long as we understand those limitations we can make wise use of the tool. Perhaps this chapter has helped your understanding of the limitations of triangular distributions. You should have a new appreciation of the importance of the minimum and maximum values; you should have an awareness of what a large y value means to the answer.

Above all, you should have a basic understanding of why base line values *alone* describe a cumulative frequency distribution.

BIBLIOGRAPHY

1. Megill, R. E., "An Introduction to Exploration Economics," The Petroleum Publishing Co., Tulsa, Okla., 1971, pg. 111, Fig. 16.
2. Newendorp, Paul D., "Decision Analysis for Petroleum Exploration," Petr. Pub. Co., Tulsa, Okla., 1975, pg. 274.
3. Lewis, E. T., "Triangular Distribution," Exxon Co. USA, Houston, Tex., memo, Feb. 25, 1972.

12 An Example of Subjective Probability

Perhaps in no other industry does subjective probability play as important a role as in the oil and gas industry. Yet its role goes unrecognized by many who employ it, derive solutions from it, or make decisions based on its use.

As shown in Chapter Nine, subjective probability goes by many names. In the search for new oil and gas fields it assumes the terms like "technical opinion" or "geologic judgment." But the principles are the same regardless of the name.

The Conditions

The simplest cases of subjective probability involve decisions where very little information exists. As more and more information becomes available the closer you move toward objective probability.

Let's set the conditions for an example of the use of subjective probability. It will be an example common to the search for oil and gas fields. It will involve subjective judgments about:

1. The presence or absence of hydrocarbons—and thus judgments about the factors for or against such occurrence;
2. Your ability to find any oil and gas fields that may exist; and
3. Whether the fields found will be economic.

When much information exists each of these three facets of subjective probability can be examined in considerable detail. In the example to follow, fortunately or unfortunately, all three facets must often be compressed into one single judgment.

Here are the conditions:

1. Four tracts are available for acquisition.
2. Very little is known about these tracts except they are on trend with other, possibly similar, tracts which produce oil and gas.

3. Several explorationists have some geologic knowledge of the general area surrounding the tracts.
4. We may or may not be bidding competitively against other companies.

The problem is to place a value on the tracts so that negotiations can begin for acquisition.

Establishing a Relative Ranking

If only one geologist or geophysicist has any knowledge about the area, then the problem is greatly simplified. One person makes all the evaluative judgments before the proposition goes to management. Since several explorationists have some general knowledge of nearby tracts they could be of help on the evaluation. We are, of course, making the assumption that "two heads are better than one." Where the available information is extremely limited this assumption can be correct—but it is an assumption. Nevertheless, in the example we shall use the judgments of all knowledgeable personnel. In the long run this action should be the wisest.

The first step is to establish a relative ranking of the properties. The ranking will be one which involves both rank *and value*. We do this because of the extremely limited amount of information. Five explorationists are given the following question:

"For the four tracts, how much from a *total value* of *10* would you assign to each of the four tracts?"

The question involves not only which is the best tract, but how much better is it than the other tracts. The answer involves both *rank* and *value*. It also allows the question to be answered in a simple manner—a subjective judgment which for each explorationist includes all of the factors he considers relevant. Each explorationist is to answer the question independently.

Suppose the five explorationists gave answers as shown on Table 12.1. Before we decide what these experts have told us, let's calculate another table. Table 12.1 gives us the exact values from the experts; Table 12.2 provides a sort of summary of the estimates. We shall concentrate on Table 12.2 but our comments will relate to both tables. Here are the judgments these estimates point toward:

TABLE 12.1

TRACT EVALUATIONS

Explorationist	Tract Number				Total Value
	1	2	3	4	
Ames	2.5	4.0	1.5	2.0	10.0
Baker	2.0	3.9	2.1	2.0	10.0
Church	2.9	4.3	0.5	2.3	10.0
Douglas	1.0	4.2	2.3	2.5	10.0
Emerson	1.9	5.5	0.7	1.9	10.0

TABLE 12.2

TRACT AVERAGE AND RANGE

Tract Number	Average Value	Range of Values	
		Range	Amt. (Δ)
1	2.06	1.0–2.9	1.9
2	4.38	3.9–5.5	1.6
3	1.42	0.5–2.3	1.8
4	2.14	1.9–2.5	0.6

1. The average ranking puts tract number two first, followed by tract four, tract one, and tract three.
2. All experts judged tract two to be the best.
3. Although tract four averaged second in ranking it was rated equal or below tract one by *four* of the evaluators. By one concept, then, tract one should be ranked ahead of tract four.
4. Tract four had the smallest range of estimates. Tracts one and three had the largest. Large ranges of values mean less certainty.

Firming the Ranking

The next step involves firming the ranking. It involves resolving the major differences so that a clearer ranking will emerge. The five explorationists are brought together to review their differences. You might wish to concentrate only on tracts one and four, since the relative rank and value of two and three should probably remain unchanged by a recycling of ideas.

What should the geological and geophysical experts try to resolve in the firming of the ranking?

1. The most obvious difficulty is tract one. Four explorationists ranked it ahead of tract four but it averaged less than four because of the ranking of one person—Douglas. His ranking and reasoning should be reviewed. A very modest change in his values between tracts one, four and three would shift tract one to second average ranking.
2. Douglas also ranked tract three, the lowest average, higher than the other four.
3. You might also want to review the differences on tract two, even though all rated it the best.

Firming the rankings may be regarded as an unnecessary step, particularly if you have very little information. After all, the real information your experts have told you is quite simple. Tract two is the best; tracts one and four are of about equal value, and tract three has the lowest value. On the other hand, firming often takes place coincident with the act of committing the money. Even after a firming of rankings, you should realize that the most diverse opinion could be right. Douglas' opinions could ultimately be the most accurate.

After a Ranking of Value

After obtaining the ranking which also gives a value concept, the next step involves the assignment of money for the acquisition. Several other considerations now enter the evaluation. These considerations include:

1. A more definite concept of field size—it may be little more than a feeling at this point, based on surrounding tracts. You already have a reflection of this in your values assigned.
2. A possible further refinement of risk.
3. The maximum value you feel can be spent on the tracts for your minimum rate of return.
4. Whether your value breaks any traditional land value concepts. You may not wish to offer $500 per acre where acreage has traditionally been only $100 per acre. Such a break with traditional values might ruin several future plays by establishing lease precedents which would put them beyond economic reasonableness.
5. A judgment as to whether your first offer will be near the owner's concept of value.

We will not go into any of these points as our main purpose has been to review a simple example of subjectively placing value on investment opportunities. The economic and strategy considerations are another story beyond the scope of this chapter. A completely new element enters if you are faced with competitive bidding rather than negotiation. Competitive bidding will be dealt with in a subsequent chapter.

What Did We Gain?

You should always consider what was gained by an exercise in subjective probability. In our example the positive aspects were:

1. If we are really interested in the tracts, we have obtained an estimate of value—on an average—from knowledgeable experts.
2. Furthermore, the answers were constructed in such a way as to give not only a relative ranking but a value reference also.
3. The relative certainties and uncertainties of the tract values show where opinions were similar or essentially undecided.

On the negative side:

1. Until the tracts are drilled, we will have no concept of validity. At this point we don't know if we have quantified ignorance or knowledge.

We do know that the best judgments have been brought to bear on an opportunity with little information and much risk.

REVIEW

Acquiring tracts of land for oil and gas exploration often involves subjective judgments. Quantifying the opinions of experts may shed light on the relative rank and value of similar tracts in an area. Decision concepts about any investment where vast uncertainty is involved benefit from gathering the knowledge of the experts.

13 Converting Uncertainty to a Cumulative Frequency Distribution

Any uncertainty that can be quantified can also be converted to a cumulative frequency distribution; and cumulative frequency distributions can be used in Monte Carlo simulation to be merged with other uncertainties to check interrelationships.

In exploring for oil and gas fields, uncertainty exists in the assessment of potential, field size and economic considerations. An economic evaluation, in reality, represents a series of forecasts merged to give an overview of an investment. In the search for new fields on the outer continental shelf (OCS) special uncertainties exist. In addition to the uncertainty associated with the presence of hydrocarbons there is the uncertainty as to sale schedule and your ability to participate in each sale. The Department of the Interior publishes a schedule of OCS sales periodically. That schedule is affected by other groups both inside and outside government. Some groups seek delay—others acceleration. Sale dates, thus, can change and even the amount of acreage offered can change. A change in the amount of acreage would change the amount of oil and gas potential offered at the sale.

Participation at a sealed bid sale is anything but certain. Large variations can and do occur for most operators in their participation from sale to sale.

In this chapter we shall demonstrate how the uncertainty of sale schedule can be quantified and converted to a distribution for further use. The method used is NOT the only method by which the distribution can be compiled—nor is it necessarily the best way. The point is not to show *THE* method, but *A* method to illustrate the conversion of an uncertainty to a distribution.

The Basic Ingredients

The model involves three basins where sales are pending. The names of these basins shall be:

The Bay of ZENITH	(BOZ)
The Gulf of APOGEE	(GOA)
The Banks of NADIR	(BON)

For simplicity's sake we shall assume that our base case is for one sale each year in each area. We desire to test the sensitivity of a one year delay or a one year speed-up of the sale schedule during a two year period. In the two year period a one year delay would remove one sale in each area. An acceleration of one year would put one additional sale into each area during the two year period.

Our reserve assessment for each sale and for the two year period in each area is:

	Millions of Bbl	
Area	Estimated Potential at Each Sale	Two Year Potential
BOZ	100	200
GOA	150	300
BON	50	100

The two year potentials are the base case in our illustration. They include the potentials for two sales, one each year. To check the sensitivity to delay and acceleration we set up the Table 13.1.

Examine the first line of this table. The base case for BOZ is 200 million bbl. If our schedule is accelerated one year, then we put one more sale in the two year period to make BOZ have an offering of 300 million bbl. Conversely, if sales are delayed one year, we have one LESS in our two-period and the exposed oil potential drops to 100 million bbl. Similar cases are set up for GOA and BON. The amount of potential per sale is the same, but the amount exposed during the

TABLE 13.1

POTENTIAL CHANGE FROM TIMING UNCERTAINTIES
TWO-YEAR PERIOD

	Millions of Bbl		
Area	One Year Early	Base	One Year Delay
BOZ	300	200	100
GOA	450	300	150
BON	150	100	50
Totals	900	600	300

two year period varies from the base case, depending upon either delay or acceleration.

As you have probably surmised, the outcomes set up the results from three different conditions. The base case contains two sales per area; three sales represents acceleration; and only one sale represents delay for each area. The example could include possibilities of a two year delay; for the sake of simplicity only, one year delays were used. We must now assign probabilities to each outcome. How probable is delay or acceleration for each area? Here, we are using our judgment as to the influence of all types of delay regardless of the cause. The cases for each area and their probabilities are shown on Table 13.2.

In Column One of Table 13.2 is shown the potential for each area under the three different outcomes. Column Two lists the probability assigned to each outcome. Column Three is the product of one times two and represents the risk-weighted potential for each outcome. Note, as always, that the probabilities for each area add up to 1.0—only one outcome can actually occur.

Column Four shows the sum of the risk-weighted values for each area. For BOZ and BON these totals are the same as for the base case. Why? The reason is that the delay and acceleration probabilities

TABLE 13.2

ASSIGNING TIMING PROBABILITIES

Area	1 Possible Potentials M̄ Bbl	2 Probability of Occurrence	3 Risk W'td Potential M̄ Bbl	4
BOZ	300	.1	30	
	200	.8	160	
	100	.1	10	
				200
GOA	450	.2	90	
	300	.4	120	
	150	.4	60	
				270
BON	150	.2	30	
	100	.6	60	
	50	.2	10	
				100
Total				570

are assigned the same value—are equal, and less than the base case. Now check area GOA. Here the risk-weighted value is less than the base shown in line two of Table 13.1. Again, why? The reason is that delay is assigned a greater probability than acceleration; therefore, in a three point probability system we are saying that the base value of 300 probably won't be achieved—but a more probable number would be less—on the average.

Once again a risk-weighted number deviates from the real world. The value of 270 million bbl is not the same as any of the potential outcomes; these are 450, 300 and 150 million bbl. The 270 is an average; and, furthermore, it is a risk-weighted average reflecting the probabilities assigned to the various outcomes. Note, however, that the risking here is for sale timing, not for variations in oil and gas potential per se.

In Chapter Nine we showed that three variables with three parameters generate 27 separate possible outcomes. Therefore, the three basins, each of which has three different possibilities, have 27 different possible combinations. As we shall see later, the 570 value for the risk-weighted sum of the nine cases is the real mean, even when we have considered the other 18 cases.

On Table 13.3 we have a rearrangement of the data to show how to generate the 27 cases. On the left are the various outcomes which can occur for each basin. Please note the column designation of a, b, and c and the row designation of 1, 2, and 3. We will be using these designations in Table 13.4 to show how the cases are derived. We have, in a sense, a matrix to determine all combinations of possible

TABLE 13.3

BUILDING 27 CASES
TWO-YEAR PERIOD

		Risk-Weighted Reserve Sizes Millions of Barrels			Probabilities of Occurence		
		Early	Base	Delay	Early	Base	Delay
	Col. →	a	b	c	a	b	c
	Row ↓						
BOZ	1	300	200	100	.1	.8	.1
GOA	2	450	300	150	.2	.4	.4
BON	3	150	100	50	.2	.6	.2
Totals		900	600	300			

TABLE 13.4

THE POSSIBLE EVENTS

	Individual Possibilities			Potential* Available	Individual Probability
1	a_1	a_2	a_3	900	.004
2	a_1	a_2	b_3	850	.012
3	a_1	a_2	c_3	800	.004
4	a_1	b_2	a_3	750	.008
5	a_1	b_2	b_3	700	.024
6	a_1	b_2	c_3	650	.008
7	a_1	c_2	a_3	600	.008
8	a_1	c_2	b_3	550	.024
9	a_1	c_2	c_3	500	.008
10	b_1	a_2	a_3	800	.032
11	b_1	a_2	b_3	750	.096
12	b_1	a_2	c_3	700	.032
13	b_1	b_2	a_3	650	.064
14	b_1	b_2	b_3	600	.192
15	b_1	b_2	c_3	550	.064
16	b_1	c_2	a_3	500	.064
17	b_1	c_2	b_3	450	.192
18	b_1	c_2	c_3	400	.064
19	c_1	a_2	a_3	700	.004
20	c_1	a_2	b_3	650	.012
21	c_1	a_2	c_3	600	.004
22	c_1	b_2	a_3	550	.008
23	c_1	b_2	b_3	500	.024
24	c_1	b_2	c_3	450	.008
25	c_1	c_2	a_3	400	.008
26	c_1	c_2	b_3	350	.024
27	c_1	c_2	c_3	300	.008
Total					1.000

*Millions of Bbl.

outcomes. First, however, note the sums of each column. The sum of column a is 900; we know this is our upper limit. No outcome can exceed 900 million bbl as this is the maximum for each basin. Column c shows a sum of 300 million. This is our lower limit. No single case can be less than 300 million bbl, as this is the least amount that will be offered in each basin under the delay case of one sale per area. These two numbers represent the extremes and all other outcomes will be somewhere between the two extremes.

On the right side of Table 13.3 are probabilities assigned to the sale schedule for each basin. The probabilities for BOZ sum to 1.0 and relate to column a, row 1, column b, row 1, and column c, row 1, (a_1, b_1 and c_1). The a_1, b_1, c_1 notation will show each specific outcome for a basin. The letters refer to columns, the numbers to rows.

Building the 27 Cases

Now that the basic ingredients are established we can proceed to build the 27 possible outcomes. Each of the 27 possibilities contains one outcome for each of the three basins. The sum of the three basins represents one possible condition of the base case, delay, or acceleration. For example, one case is the sum of a_1, a_2, and a_3 or 900. This is our maximum case. This case has a low individual probability, however, ($.1 \times .2 \times .2 = .004$). It would occur *on the average* only once in 250 events.

Another possible case from Table 13.3 would be ($c_1 + c_2 + a_3$) or 400 million bbl. It has a higher individual probability ($.1 \times .4 \times .2 = .008$). It would occur only once in 125 times, on the average. Note, we have used the term *individual* probability. In a moment, we shall see that more than one possible outcome produces an answer of 400 million bbl.

All 27 possibilities are listed in Table 13.4. The method for their calculation can be followed from the letter and number code under the column headed Individual Possibilities. Line 1 represents the first possibility which we illustrated before—a_1, a_2, a_3. The next shown, line 2, is a_1, a_2, b_3. Follow these letter-number designations and you can sum the three basins to produce all 27 possibilities.

The reserve sums are shown under the column headed "Potential Available." Note, carefully, that several combinations produce identical values—just as several combinations of two dice can produce a seven!

The last column shows the *individual* probabilities for each of the 27 possibilities. Again, note that the sum of all probabilities is 1.000.

What do we do about the possibilities which give identical values? We make a distribution!

The Making of a Distribution

On Table 13.5 we build a distribution. Following rules laid down in Chapter One class intervals are established and for cumulative purposes, later, the individual cases are arranged in order, the largest first.

The rearrangement of the data shows interesting facts.

TABLE 13.5

RANKING OF POSSIBLE OUTCOMES

		Probability		
No.	Possible Outcomes	Individual	Grouped	Cumulative Grouped
1	900	.004	.004	.004
2	850	.012	.012	.016
3	800	.004	.036	.052
4	800	.032		
5	750	.096	.104	.156
6	750	.008		
7	700	.024	.060	.216
8	700	.032		
9	700	.004		
10	650	.008	.084	.300
11	650	.064		
12	650	.012		
13	600	.008	.204	.504
14	600	.192		
15	600	.004		
16	550	.024	.096	.600
17	550	.064		
18	550	.008		
19	500	.008	.096	.696
20	500	.064		
21	500	.024		
22	450	.192	.200	.896
23	450	.008		
24	400	.064	.072	.968
25	400	.008		
26	350	.024	.024	.992
27	300	.008	.008	1.000

1. The two highest and two lowest values are represented only *once*.
2. All others occur two or three times, but never more than three.

We could make a distribution from the first column; *if* we considered each possibility equally likely, it would fairly represent our expected frequency. Remember, however, that we do *not* consider each outcome equally likely. We are definitely interested also in the probabilities of each possible outcome.

The second column on Table 13.5 lists the *individual* probabilities for each outcome. The third column groups the individual probabilities for the identical outcomes. For example, three possible combinations can produce 600 million bbl. The sum of their probabilities (.008 + .192 + .004 = .204) represents the probability of the occurrence of 600 million bbl. This value, .204, represents the highest probability for any value but is only slightly higher than the value of 450 million bbl. In like manner the individual probabilities are summed to give the combined probability for identical values.

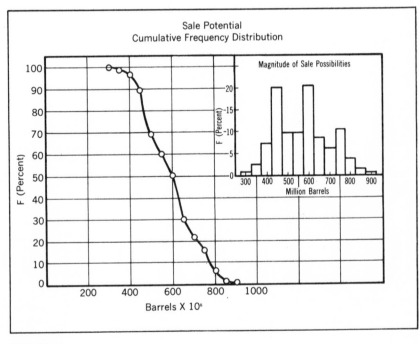

FIG. 13.1

Column 4 shows cumulative probability for the values, to be used in constructing a frequency diagram.

The diagram is Fig. 13.1. This diagram has two parts. On the left is a "greater than" cumulative probability distribution which was constructed from the frequency distribution on the right.

Let's begin with the frequency distribution. It hardly represents any of the ideal distributions reviewed previously. It is essentially bimodal with the two peaks at 600 million bbl and 450 million bbl. A glance at the frequency distribution seems to indicate that delay (smaller values than 600) is more likely than acceleration (values higher than base of 600). The mean of our distribution is 570 million bbl, indicating that delay does have a slight edge. Remember we introduced this delay in GOA by saying that delay had a greater probability than acceleration. In the other two basins the base was the most probable outcome.

The cumulative frequency distribution on Fig. 13.1 is the culmination of the exercise. We have converted opinions about an uncertain set of conditions into a cumulative frequency distribution. This distribution can now be entered with other uncertainties in the form of distributions to bring to bear on the desired answer, all of the influences affecting the possible outcomes. We have quantified our opinion—and in usable form.

REVIEW

Opinions can be quantified into usable forms to shed light on factors affecting possible outcomes. Most often the desired final form is a distribution. Not all distributions will be normal, lognormal or even smoothly continuous. They will, however, reflect opinion. We state opinions about each parameter and sum toward the final answer. As we shall see later, this step is one of the fundamentals of risk analysis.

14 Basin Assessment

If it happens, you must admit it was possible!

The assessment of oil and gas potential is one of the integral parts of an exploratory program. An intuitive and subjective art in recent years, basin assessment today receives much more attention and concern than in previous years. The reason is simple. Our nation has become totally dependent on foreign supplies for over 40% of its daily usage of oil. The Federal Government in particular has recognized the importance of knowing how much oil and gas remains in unexplored areas of the outer continental shelf and the inland areas of the United States. At present and for the next decade, we have no immediately available alternate to oil and gas. It becomes extremely vital, therefore, to know what our remaining storehouse holds in the way of undiscovered oil and gas fields. However, we do not know what is in this storehouse. We can only estimate what exists with great uncertainty as to our answer; so much uncertainty is involved that we express our answer as a probability distribution, showing the full range of answers which might occur.

Thus, basin assessment is an exercise in risk analysis. In this chapter we shall review briefly the answers of published assessments and several methods for making basin assessments.

Current Assessments

Numerous authors have been interested in the remaining potential for oil and gas fields in the United States. Table 14.1 shows recent estimates of various individuals or organizations on remaining potential. These estimates range all the way from Hubbert[1,2] on the low side to high side estimates by Exxon, line 5. The Potential Gas Committee[3] has an even higher estimate than Exxon, but it is not shown and is less up-to-date. In between are various estimates by companies and individuals. Methods used are not similar and a few estimates (North,[4] not shown on Table 14.1) represent judgments about other existing assessments. The low estimate by Exxon, line 1, Table 14.1, is the

TABLE 14.1

MEAN ESTIMATES OF TOTAL UNDISCOVERED LIQUIDS AND NATURAL GAS—U.S.A.

Line	Source	Date	Amount BB*	Amount Tcf*
1	Exxon[a]	1976	63	287
2	Hubbert	1974	72	480
3	Mobil[b]	1974	88	443
4	USGS	1975	98	484
5	Exxon[b]	1976	118	582
6	Shell	1975	110	400
7	Nat'l Acad. Sc.	1975	113	530

*Billions of barrels of crude oil (plus condensate and natural gas liquids) or trillions of cubic feet.

[a] Attainable portion of resource base, line 5.

[b] To 6,000-ft water depth; other estimates to about 600 feet.

amount technologically and economically attainable from the larger number, line 5, called a resource base.

Each individual company or organization, however, faces the same perplexing problem—thinking about the unknowable! Gradually those outside the petroleum industry are beginning to understand that in thinking about the unknowable, there is much room for both difference of opinion and error. Remember the knowledge curve (Fig. 9.1).

The subject of basin assessment will be the object of many methological attempts in the years to come. It will receive increasing attention of both government and academia. The truth is, however, that basin assessment has been around a long time. It has been only in the past few years that the methodology used by knowledgeable groups has been published.

In the area of basin assessment we can expect rapid evolution of published ideas, because of the vital importance of this subject to our nation. The fundamentals will remain the same, however, and it is the fundamentals we will discuss.

An Assessment Curve

Before illustrating an assessment method used in Canada, let's first review an assessment curve. In Fig. 14.1 curve *a* represents a typical assessment curve. Numerous such curves are in the United States Geological Survey's (USGS) latest estimate of undiscovered recoverable resources.[5]

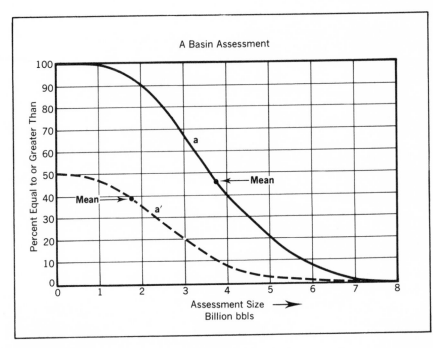

FIG. 14.1

Curve *a* represents a basin with established production and, thus, has a 100% chance of at least 1 billion bbl. Note there is almost no chance of the basin containing more than 7 billion bbl. The mean of this distribution is 3.75 billion bbl at a 45% chance of the assessment being equal to or greater than this amount. (See Appendix B for a simple method of calculating the mean of any distribution).

Curve *a'* represents an untested basin with a 50% chance that it contains no hydrocarbons at all. One can state this fact another way. There is only a 50% chance of finding anything! The mean of Curve *a'* is only about 1.7 billion bbl because of the large chance of finding nothing.

Finally, if each value on *a'* were one-half that of *a*, then *a'* would represent Curve *a* risked at 50%. In such a case the mean of *a'* would be one-half the mean of Curve *a*.

An assessment curve shows valuable information about a basin.

1. It shows the full spectrum of expected results, a minimum and a maximum concept.

2. It can express the chance of a basin being totally dry—the chance of finding anything.

With this reminder of the use of a cumulative frequency distribution, we are ready to illustrate an assessment concept.

The Delphi Approach

One of the earliest methods used in risk analysis has an unusual name—the Delphi technique. The name Delphi is of interest in itself as a choice for this technique. Any student of Greek mythology would recognize Delphi as the ancient city of Greece which was the site of the celebrated oracle of Apollo. Anything delphic is, therefore, oracular, ambiguous and obscure in meaning. It would be well to remember this description. Although the Delphi technique will be illustrated, primarily because it is still used in some instances as an overall technique, it is not recommended as the best method. Its flaws will be obvious to the reader as we review the Delphi method.

The Method

The following method follows closely one outlined in detail in "An Energy Policy for Canada," Chapter Two, pages 31-40.[6] This reference will be reviewed subsequently.

Delphi techniques can begin at various states of an assessment. They can also be used as a means of arriving at values for key variables. In its earliest usage it was similar to a brainstorming session in that an overall answer was desired and estimated from the technique with essentially no intermediary steps.

Suppose we wish an assessment of the South Sea basin. It is unexplored; no wells have been drilled; our seismic data is sketchy, limited and of questionable value. We need a quick estimate of oil and gas potential. So, we gather several knowledgeable explorationists in a room and pose questions. We are really applying our skill to extract opinions from the experts as described in Chapter Nine. Everyone in the room recognizes the limit of real knowledge about the South Sea basin.

After reviewing the "knowledge" which is available from geology, geophysics, etc., we pose the following questions to our panel of experts.

1. What is the probability of something being discovered?

2. What is the largest quantity of oil and gas you can conceive as being discovered from this basin?

If we stopped with these two questions we would in reality have the end points of a cumulative probability distribution. Question one defines the 100% point of a "greater than" curve and question two the zero percent point of the distribution.

The Canadian paper recognized the possibility of asking for intermediate points. We could better define our curves with a request for five points. These questions could be asked in the form of a brief table.

Probability of Reserves Being Greater than Stated	Basin Potential (Billion Barrels)
0	—
.25	—
.50	—
.75	—
1.00	—

Let's examine each point on our curve.

1. The zero point represents the maximum possible estimate. It is question two in the preceding illustration. You are saying there is no chance of the basin being larger than this amount.
2. The .25 probability means you feel there is only a 25% chance of ultimate reserves being greater than this number.
3. The .50 point is an approximate "most likely" estimate. It is a median. One-half of your estimates are above this amount and one-half are below.
4. The .75 probability indicates you think there is at least a 75% chance that the basin contains reserves exceeding this amount.

TABLE 14.2

PROBABILITY RESERVE ESTIMATES SOUTH SEA BASIN—BILLION BARRELS

Probability	Expert				
	1	2	3	4	5
0	20	40	30	10	60
.25	15	20	20	6	35
.50	10	7	15	4	17
.75	5	2	10	2	10
1.00	2	0	5	0	0

5. The 1.00 probability is your low estimate. It would reflect question one of the previous illustration. It could also be a number less than 100% if you think there is some chance of the basin being completely dry. On the other hand, for a basin with a few productive fields the 100% point represents the absolute certainty of what has already been found.

A poll of the experts yields the following results:

The five points (0, 25, 50, 75, and 100) from each expert are plotted on Fig. 14.2. The individual curves have the participant number near the bottom end. Here we see the marked individuality of the estimates. For example:

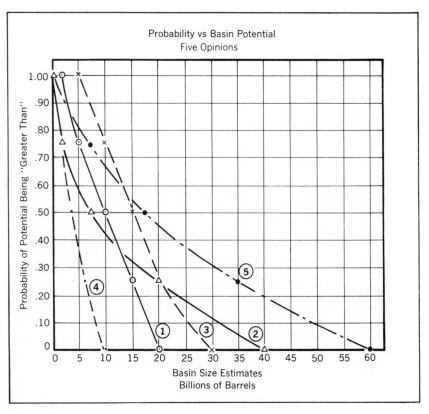

FIG. 14.2

1. Participant four is the most conservative. He sees no chance of the basin exceeding 10 billion bbl.
2. On the other hand, participant five is the most optimistic. He estimates the basin's upper limit at 60 billion bbl.
3. Note the shape of the individual curves. Participants two and five see some probability of very small potential. They also see better probabilities of a high potential than the other participants.

You may see other bits of information from the individual curves. The next step is the calculation of an average or composite curve. Such an average is often compiled by a computer. In this particular application, where all curves describe the same quantity (barrels) we can manually calculate, with accuracy, an average curve.

A Composite Curve

The manner in which a composite curve is calculated is illustrated in Table 14.3. On the left most column are shown even increments of billions of bbl. To the right are values of probability (percent in this case) for each participant for this barrel number. We, thus, *average the probabilities.*

Why not average the reserve estimates at a given probability—i.e., reading horizontally, the five reserve estimates for each five billion barrel intervals? The reason is simple. This method would not produce the

TABLE 14.3

PROBABILITIES (PERCENT)—SOUTH SEA BASIN

Billions of Barrels	Participants					Arithmetic Mean
	1	2	3	4	5	
5	75	58	100	35	85	71
10	50	42	75	0	67	47
15	25	32	50		55	32
20	0	25	25		45	19
25		18	12		37	13
30		12	0		31	9
35		6			25	6
40		0			20	4
45					18	3
50					8	2
55					4	1
60					0	0

right average, particularly on the high side. If you average the zero probability points of 10, 20, 30, 40 and 60 billion bbl you get 32 billion bbl as the maximum for the South Sea basin. Such an answer is illogical as two of the participants, two and five, have estimated the maximum considerably higher than this. By averaging the percentage estimates you show a probability, though small, of high side estimates up to 60 billion bbl.

The arithmetic mean of the percentage estimates from Table 14.3 are plotted on Fig. 14.3. Our composite shows:

1. Only a 70% chance of the South Sea basin having oil reserves greater than 5 billion bbl.
2. It estimates a 50% chance of as much as 9.5 billion bbl.

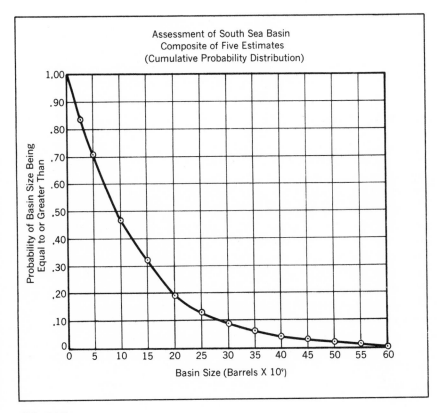

FIG. 14.3

3. A ten percent chance is indicated for 28 billion bbl and a 5% chance for 37.5 billion bbl.

Our composite curve is skewed sharply to the right with small chance of a high side potential several fold the median value of almost 10 billion bbl. The mean of our distribution is about 13 billion bbl, reflecting the skewness of our curve. There is only a 38% chance of the basin achieving the mean value or more!

What Now?

Fig. 14.3 represents our view of the potential for hydrocarbons in the South Sea basin. It is the composite of five opinions, showing in average the complete spectrum of possibilities. We could end the exercise here and go on to the next project.

However, one or two additional steps may be desired. Suppose you feel the distribution is skewed too far to the left; or perhaps you think the low side estimates (95% probability) discount the basin too heavily. You can reconvene the group and discuss each input point—but only if you feel there are sufficient data to warrant an additional session.

The Flaws of Delphi

The Delphi technique of brainstorming can often produce answers surprisingly close to reality, but it may also produce one not even in the "ball park." For explorationists the basic flaw of Delphi is its emphasis on the answer rather than the pieces which make up the answer.

Consider our example of the estimates of field size, Chapter Nine. One can survey the experts to get a total prospect size; but a better way is to use questions about ranges of pay thickness, areal extent and recovery per acre. Triangular distributions are recommended here with their three values of minimum, maximum, and most likely. These can then be combined to produce a complete spectrum of answers—based on the reality of measurable and understandable quantities.

So, a better use of the Delphi technique is to gather opinions about the ingredients which make up the final estimate. Place your probabilities or distributions on the basic factors and a more realistic answer should result.

A fundamental of risk analysis is to Delphi the parts and sum to the whole.

A second major problem of Delphi is that for any basin it can ignore important facts. Geologic considerations of source, trap, environment, and structure might be individually risked as shown in Chapter Nine, to determine the likelihood of any production whatsoever.

Other equally important factors involve the number of known structures from seismic data and the expected field size distribution. The 1975 USGS estimates of undiscovered oil and gas resources recognize the significance of lognormality and proper field size distributions.[5] These are important considerations in good basin assessments. The Canadian report[6] also discusses field size distributions—but briefly.

Other Methods

In a critique of the USGS pre-1975 estimates, the National Academy of Sciences[7] lists several methods of constructing basin assessments. Various methods are also listed in the USGS Circular 725.[5] Nanz[8] also lists methods for assessing basins. Haun[12] briefly described methods presented at a research conference sponsored by the American Association of Petroleum Geologists (AAPG), presented at Stanford University in August, 1974.

In general these methods can be grouped as follows:

I. Volumetric Methods.   A volumetric estimate begins with the physical volume of sedimentary rock in a basin or play. In a well explored basin with many wells drilled volumes can be estimated with reasonable accuracy. In a frontier basin volumes have less accuracy but can be estimated from geological and geophysical data—the latter largely seismic data.

Next a recovery per cubic mile of sediment is related to basin volume to estimate total recovery.

$$\text{Volume of sediments} \times \text{Bbl per unit of sediments} = \text{Bbl}$$

This two step approach is simplicity at its finest. However, it is loaded with flaws. These are:

1. The most obvious flaw involves the choice of the right recovery per unit volume. The possible answers here are numerous. One can:
    a. use the world wide average from productive basins
    b. choose a similar basin and apply its recovery
    c. estimate a recovery based on judgments from a and b.
2. If *1a* is chosen it may have little relevance to the real situation. In the absence of any better logic, you may wish to use it, but

you should do so with your eyes open. The world wide mean recovery is an average from many possible answers.

3. If *1b* is applied, you are saying you can pick a similar basin—Jones[9] argues that each basin is unique, and that there is a lack of correlation between geologic factors and recovery per unit volume. He suggests an estimation method based on four factors:

   a. source
   b. trap
   c. reservoir
   d. migration

Each factor is important in the total estimate.

Today few knowledgeable assessors advocate the volumetric method. It is a method of last resort which ignores too many critical geological, geophysical, and economic parameters.

II. Mathematical Methods. Various estimators have used mathematical models to estimate future undiscovered resources. Hubbert's[1,2] studies are the most famous. In the late 1950's and early 1960's he was using a roughly bell-shaped curve to illustrate the production of a finite resource. The validity of Hubbert's methods for a continuously developing area is not questioned. Their application to areas not yet developed or having sparodic development leave considerable doubt. Some production must occur before the shape of Hubbert's curve is known. Thus, the validity of Hubbert's method to the unknown, untested basins in the outer continental shelf is questionable.

White, et al,[10] have used extrapolation of existing trends in mature basins to predict future discoveries. Extrapolations of historic data (Marsh)[11] establish the growth of existing discoveries. Existing discoveries in turn are extrapolated to an economic limit to establish estimates of future discoveries. The extrapolated units involve some measure of discovered reserves. If drilling levels and average depth per well were constant the average discovery size could be extrapolated. Neither situation exists in the real world so most users of this technique use barrels found per foot of exploratory drilling plotted versus cumulative barrels found. The value of cumulative barrels as one factor is its reduction of the element of time as an overpowering factor. Time is not completely eliminated as the barrels found per foot drilled are an annual factor.

Extrapolations also have their problems, but these are by degree much less serious than the volumetric base. For example:

1. On the plus side, you are relating to history as you see it; but

negatively, you are assuming that history provides a sure key to the future.

2. Changing economic conditions can affect your economic limit by influencing the amount of drilling to be done or the recovery from the reservoirs in new oil and gas fields found.

3. Several forms of extrapolation exist—linear, logarithmic, etc. You must choose one which you deem appropriate.

4. Extrapolations must have a stable broad base of historic data. Lacking this strong statistical base they are valueless. Thus, extrapolations are of no value in frontier areas. There is no history to extrapolate.

White, et al.,[10] recommend a method of extrapolation developed by Baker.[10] It produces multiple answers which can be converted into a cumulative probability distribution. Such output offers the advantage of allowing its combination with other distributions.

III. Geological Methods. Geological methods comprise the last general group of assessment techniques. These methods are the most popular for frontier areas, if the data base is adequate. An inadequate data base pushes many estimators toward the volumetric methods.

Geologic methods take advantage of all geologic data, all geophysical data and environmental and depositional conditions. They can include (Nanz)[8] an estimate of the number of traps for oil and gas which exist in the frontier. To be sure, judgments are involved. However, geologic methods allow the best consideration of the most significant unknown in frontier exploration—namely "Do oil and gas fields exist in this basin?"

Frontier areas, basins or plays, usually involve either completely untested areas or untested sections of sedimentary rocks. In the latter case, it is usually the deeper rocks which remain untested. A deep unexplored section of sedimentary formation can have just as many unknowns and just as many risks as a new basin in which no well has ever been drilled.

In Chapter Nine, on risking prospects, we touched on the problems associated with answering the question "Is it there?" The same set of questions are involved in basin assessment. We might consider only the three factors of structure, reservoirs and environment. Estimating the value of these factors for basin assessment necessitates judgments. Here the factors assume a broader scope, however. For example—it is not a question of a single structure but in the absence of geophysical data a question of whether structures do exist. The factors of reservoir and environment are equally broad. Was the oil and gas generated in

economic quantities? Did it migrate to potential traps after the traps
were formed? If so, has the oil or gas remained at the site of the trap?

In basin assessment, as in risking prospects, risk factors are assigned
to each key variable. Uncertainty is thus expressed in numerical form
and, applied to barrel estimates, can produce a risk-weighted estimate
of hydrocarbons for an untested frontier basin.

As with a pay off table showing expected value, in basin assessment
a risk-weighted basin assessment is not a real world number. It is only
one of many possible answers and takes into account your estimates
of the chances of this and other similar basins being dry. For this reason,
basin assessment is most frequently expressed finally as a cumulative
frequency distribution. Only from the composite of many such risked
distributions can one estimate with reasonableness the range in quantities
of oil and gas to be found.

In summary, geologic basin assessment places value judgments on
key parameters which are largely unknown. Either through volumetric
estimates, field size distributions, or counts of known prospects, these
judgments are applied to arrive at barrel estimates. The final product
is most often a cumulative frequency distribution. In a few paragraphs
such a distribution will be shown and discussed.

IV. Combinations. Combinations of I, II and III have been used.
White, et al, used both extrapolation and a volumetric method for an
estimate of oil and gas potential for southern Louisiana. Geologic and
volumetric methods are sometimes combined for untested basins.

What determines the method used? It depends partly on the estimator
and his experience, but it is more directly related to the type and quantity
of data available. Most estimators would agree with Nanz that methods
based on geology are the most reasonable and reliable, particularly for
basins with some seismic information available. In a basin where no
wells have been drilled and no seismic data are available, any method
has limits. From the knowledge curve, Chapter Nine, we learned that
the less that is factual the greater the range of possible opinion; and
whether we like it or not, assessments under conditions of limited data
are largely opinion. It should be expert opinion, but it remains opinion.

Reality Check

One practice every good assessor adopts quickly, whenever data
are sparse, is the reality check. The easiest way to accomplish a reality
check is to make two estimates of basin potential using two different
methods. The answers should at least be similar, if not, check your
input to each method and recalculate your answers. Relate your estimate

to a similar basin for a reality check. Such a check assumes you believe a similar basin exists. Check your reasons for assuming similarity. They should involve both the geologic setting, framework and field sizes.

Increase your objectivity by assessing several basins at one time. The experienced estimator has the benefit of past assessments as a guide. One learns assessing by doing, as with most techniques, and making more than one assessment will increase your objectivity and the accuracy of your results.

Finally, remember that you are dealing with many factors which at the moment of your assessment are unknown. Your hope is that you make the best possible estimate from available data; but recognize that especially when the facts are few, another might take the same data and arrive at another somewhat different conclusion.

Attainable Potential

An estimate of total resource potential must often be corrected for reserves which remain unattainable. Marsh[13] describes attainable potential as "the amount of recoverable hydrocarbons in reservoirs and accumulations whose productivity and size make their development economically feasible in the foreseeable future." Reserves to be attainable must be producible within the scope of present or imminent technology. Thus, the reasons some potential is not attainable are:

1. Field sizes too small to be economic.
2. Ultra deep water, which raises costs, stretches technology and rules out small accumulations of oil and gas.
3. Hostile environments (massive icebergs, for example).
4. In the OCS fields too far from shore can result in poor economics especially for gas fields.
5. Political and economic factors (laws, wild life refuges, etc.) can eliminate some otherwise attainable potential.

A necessary step, then, in basin assessment is to adjust, downward, your total potential to that which is attainable. Such reductions can be judgmental as well as geological and economic. Yet the adjustment is necessary. You never bid on potential that cannot be obtained economically with existing or soon-to-be-developed technology.

Small Fields and Attainable Technology

In any trend there exists a lower limit to the size of fields which can be developed economically. Even at shallow depths there is some field size that is so small, profits will not result. This field size represents the economic cutoff for any field size distribution. Fields smaller than the cutoff size are non-commercial and their discovery wells are classed as dry holes.

Granted, we produce a few small fields which never pay out, just to get part of our money back; but you would not deliberately look for a field below your economic cutoff size.

Therefore, in using a field size distribution (FSD) for play assessment, the economic cutoff size (ECS) is an important parameter.

Field Size Distributions (FSD's) and Success Rate

We have dealt at length in several chapters on the meaning of success rate. As an average, it represents the chance of finding anything—or the chance of finding the smallest size field or larger.

The economic cutoff size (ECS) is the smallest size that will support your play and the average success rate relates to this field size or larger. The ECS is illustrated by line aa' on Fig. 14.4. Any field smaller than this size is considered uneconomic.

Let's assume in the trend under examination that the average success rate, for fields larger than the ECS, is 20%. What happens to the success rate if we change our mind about the economic cutoff size? Suppose a crude price increase permits the search for smaller fields. If so, the total number of potential fields has increased because we have lowered our ECS. Correspondingly, we have raised our chances of success—since there are more fields to find we have expanded the possibilities for success. We may not have added many barrels but we have increased the number of potential finds. Suppose the ECS is lowered to bb' as shown on Fig. 14.4. The new success rate can be calculated using the percentages (at ECS) for the two lines aa' and bb' as follows:

$$\frac{\text{new ECS cutoff } \%}{\text{old ECS cutoff } \%} \times \text{success rate (old ECS)} = \text{new ECS success rate}$$

For bb' then the new success rate is:

$$\frac{85.5}{69.0} \times 20 = 24.8 \quad \text{or almost } 25\%$$

A crude price cut would eliminate some fields by making them uneconomic. The decrease in the number of potential fields raises our ECS and lowers our chance of success. Line cc' on Fig. 14.4 illustrates

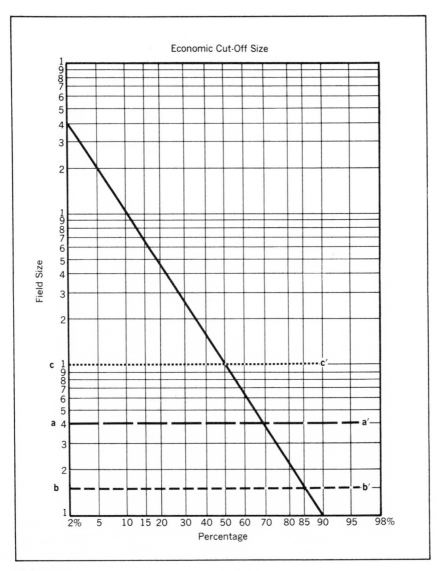

FIG. 14.4

a higher ECS, fewer fields and reflects a lower success rate for the reduced number of fields, calculated as follows:

$$\frac{50.0}{69.0} \times 20 = 14.5\%$$

Our original discussions of a FSD demonstrated the rareness of the largest fields in any trend or play. The chances of finding the large fields are less, simply because they are in the minority in your distribution. Raising the ECS, then, lowers your chance of success in a trend or play. In other words, your chances of failure increase as you raise the ECS. If you are successful, however, your discoveries will be among the largest; your success rate will be affected more than total barrels found.

FSD's, Success Rate, and the Real World

The foregoing abstract discussion of the relationship between ECS and success rate must be qualified to some extent. Under ideal circumstances a company has been able, occassionally, to lease an entire basin. A foreign concession could represent such a circumstance. Under these conditions your prospects could be listed by size and the largest drilled first. You could conceivably find the largest one first!

The concept and use of a FSD for assessment purposes does not rule out this favorable circumstance! The important point to remember is this: large fields are in the minority in every trend. Therefore, your chances of finding the more numerous small ones is larger. Even if you drill them in a pre-drill sequence of expected size, you can use the concept of a FSD:

1. to estimate your ECS
2. to gain an understanding of expected size
3. to relate to the trend average success rate to illustrate the chances of finding a specific field size.

REVIEW

Basin assessment is attempting to define quantitatively the unknown and possibly the unknowable. Yet it must be done—it is done by all companies planning investments in untested and semi-tested basins as well as by agencies in the Federal Government. The techniques are slowly being exposed to the printed page—but with an increasing rate. The answers will always, however, be approximate for an untested basin. Only the drill can fully answer the greatest unknown—"Is it there?"

Basin assessment involves:

1. Gathering all the facts available
2. Choosing a method or combination of methods to make the best assessment
3. Applying the method
4. Making reality checks.

BIBLIOGRAPHY

1. Hubbert, M. K., "Degree of Advancement of Petroleum Exploration in the U.S.," AAPG Bulletin, V. 51, 1967, pgs. 2207-2227.
2. Hubbert, M. K., "U.S. Energy Resources, a Review as of 1972, Part 1, 1974," Interior Committee Documents, Senate Office Bldg., Washington, D.C.
3. Potential Gas Committee, "Potential Supply of Natural Gas to the U.S.," Potential Gas Agency, Colorado School of Mines, Golden, Colorado, 1973.
4. North, F. A., "A Sane Look at U.S. Gas Resources," National Gas Survey, Volume V, F.P.C., 1973, pg. 113.
5. Miller, B. M., Thomson, H. L., et al, "Geological Estimates of Undiscovered Recoverable Oil and Gas Resources in the U.S.," Geological Survey Circular 725, July, 1975.
6. "Energy Reserves and Potential Resources of Oil & Gas," An Energy Policy for Canada, Phase I, Volume II, Chapter 2, pg. 31ff.
7. Gillette, Robert, "Oil and Gas Resources; Did the U.S.G.S. Gush Too High," Science Vol. 185, July 12, 1974, pg. 127.
8. Nanz, R. H., "The Offshore Imperative—the Need for and Potential of Offshore Exploration," Presented at Colloquium on Conventional Energy Sources and the Environment, University of Delaware, Newark, Delaware, April 30, 1975.
9. Jones, R. W., "A Quantitative Geologic Approach to Prediction of Petroleum Resources," from "Methods of Estimating the Volume of Undiscovered Oil & Gas," AAPG Studies in Geology No. 1, Tulsa, Okla., 1975, pg. 186.
10. White, D. A., et al, "Assessing Regional Oil and Gas Potential," from "Methods of Estimating the Volume of Undiscovered Oil and Gas," AAPG Studies in Geology No. 1, Tulsa, Okla., 1975, pg. 143.
11. Marsh, G. Rogge, "How Much Oil Are We Really Finding," Oil and Gas Journal, April 5, 1971, pgs. 100-104.
12. Haun, J. D., "Methods of Estimating the Volume of Undiscovered Oil and Gas," AAPG Studies in Geology No. 1, Tulsa, Okla., 1975, pg. 1.
13. Marsh, G. Rogge, "Forecasting Supply and the Timing of Production in Offshore Frontiers," in Davis, J. C., Doveton, J. H., and Harbaugh, J. W., Convenors, Probability Methods in Oil Exploration—Amer. Assoc. of Petroleum Geologists, Research Symposium, Aug. 20-22, Stanford University Preliminary Report, 1975.

15 Competitive Bidding

All of the elements of risk analysis come together in a competitive bid for oil or gas leases on the outer continental shelf (OCS). There are five basic questions which must be answered to arrive at a bid. They are:

1. Are oil and gas present. (Is it there?)
2. How large is the deposit. (How big is it?)
3. What profits are possible. (Will it make money?)
4. What are reasonable bids. (How much can I afford to pay for it?)
5. What competition is possible. (What are the other guys going to bid?)

Competitive bidding exists in many industries, not just the oil industry. Many government contracts require competitive bidding. Furthermore, published literature on this subject is quite extensive. For example, Stark[1] published in 1971 a list of over 100 papers and books dealing with competitive bidding.

In this chapter our concern will be with competitive bidding as it applies to the oil and gas industry. We shall do this by first reviewing significant papers published on the subject in recent years and conclude by summarizing what these works "say" in composite.

Bidding Patterns

A good place to start is with Paul Crawford's paper on Texas offshore bidding patterns.[2] In reviewing his work you will understand why we performed some of the exercises in Part I, Chapter Four. Particular reference will be made to the multiplication of the face values from rolls of the four dice to achieve a lognormal distribution.

Crawford begins with the logical premise that a final bid results from the multiplication of several factors, each of which exhibits uncertainty. Their product is a single answer. But the uncertainty in each factor (pay thickness, area, porosity, risk factor, etc.) is akin to one set of faces on the four dice. Another roll produces four different faces. Another competitor produces four different values for the same factors.

Thus, a study of competitive bidding begins with the important fact that single bids by different competitors should arrange themselves into a reasonable facsimile of a lognormal distribution.

Crawford includes plots of bids on six tracts, five of which show an approximately lognormal distribution.

Another key point comes from this paper. It stems from the number of factors considered. The greater the number of factors used to arrive at the product (bid) the greater will be the variance. The spread can be large between bidders if:

1. the variance is 20% to 30% above or below the average, and
2. the number of factors multiplied to arrive at the bid is three or more.

An example of the very large range possible is Tract 230 in the Texas offshore sale of 1968. The ratio of the high bid to the low bid was 100:1.

In summary, we begin with two important facts. Bids on a single tract will be lognormal and the range of bids can be very large.

Next—A Bidding Model

We can now proceed to a bidding model. Our concern will center on what the model reveals, not its construction. Anyone interested in the latter can review thoroughly the reference article. It is the work by Capen, et al.,[3] on competitive bidding in high risk situations using a simulation model. The model simulates the bidding game.

First let's trace briefly what is involved in any bidding process. It all begins with an estimate of field size which is then translated to values of reserves per tract. From reserves ensues a schedule of production which converts easily to revenue; with the addition of exploration and development investment plus approximate costs, we get cash flow and net profit. From risk and return concepts we translate profit into what we are willing to bid.

This process is in one sense nothing more than a typical economic evaluation of the type illustrated in Exploration Economics.[4] However, a major new factor enters the evaluation in the form of competition. You must address all of the factors normally included in an economic evaluation *plus* the consideration of what the other bidders might be doing.

Competitive bidding for oil and gas leases almost always involves great uncertainty and greater opportunity for loss than competitive bidding in other business activities. Granted there are uncertainties in a bid for constructing a building, laying a pipeline, or obtaining a fuel contract; the difference is one of degree. You can more easily lose your entire investment bidding on an oil and gas lease. Such a loss is much less common in other types of competitive bidding.

From prior chapters we have seen the variance possible in an estimate of field size. Reserve size and the estimation of risk are the most critical factors in competitive bidding. From the considerations on reserve size comes the first key point from the simulation model used by Capen, Clapp and Campbell. There is much uncertainty in the factors resulting in estimates of field size and large variations possible in risk estimates. Therefore, the bidder who wins will most often be the one who overestimates the reserves on the tract. This key result occurs because of the nature of competitive bidding. You tend to lose the tracts you correctly estimate—and win those you overestimate. The result is that your winnings—the high side reserve estimates—tend to produce only modest profits—not what you hoped to average at a given sale. Capen, et al., point out that your bids, on the average, may indeed produce desired profit goals but you don't win bids based on your average—but only on the highest reserve or value estimates.

To effect a better strategy, Capen, et al., recommend lower bids when:

1. you have poorer information than competitors
2. you are less certain, than usual, about your own size or value estimate
3. you expect more than three bidders.

The last recommendation stems from the frequent occurrence of many bidders on the good tracts. The larger the number of bidders the higher the winning bid, in general, and the better the chance of only modest profits, if any. Capen's model would steer the bidder away from the most popular tracts—in effect—by purposely submitting a lower than normal bid.

You might consider this rationale very defensive. Capen, et al., defend this concept by showing the broad range which often exists between the highest (winning) bid and the second highest. That range, they say, leaves room for lower bids and still a reasonable chance to acquire properties. They do not deny that such a strategy will result in fewer tracts acquired; but they want to optimize profits, not just acreage acquisition.

The Winner's Curse

We might leave this excellent paper at this point—but it has much more to offer. The authors present an example illustrating why more bidders push you toward lower returns—by dropping out your lower underestimated tracts. This tendency they nicknamed "the winner's curse."

From these discussions two more key factors emerge:

1. The chance of winning is more related to the estimate of reserve size than upon a particular bid level. Putting this in different words, modest increases in minimum rate of return do not drastically reduce your chance of winning acreage.
2. Because of the significance of the estimate of reserves, the greatest possibility for errors in reserve size exists in frontier areas.

Capen, et al., conclude their article with comments about entering their model with factors reflecting a competitor's prior bidding patterns. Overall their paper has been one of the most important about the subject of competitive bidding to be published in recent years. It is recommended reading for all who want to learn more about the fascinating game of bidding and winning without "losing your shirt."

Their principal point is of prime significance. Every bidder, on the average, evaluates his properties adequately; he will win, however, not on the average, but when he overestimates value the most. This unique characteristic of competitive bidding pushes individual companies' returns below their expectations. The tendency to low returns is not the fault of the evaluators—the explorationists—but a function of the competitive bidding system itself.

Bidding and Producing Relationships

Lohrenz and Oden[5] analyzed the relationships between bidding levels and the subsequent amounts of production from leases acquired. Their analysis showed conclusions similar to those expected from the simulation model of Capen, et al., and subsequent models—to be reviewed in this chapter. They analyzed leases sold up to and including the March, 1962, sales.

Their principal conclusions were:

1. As the number of bids increased per tract the bonus increased as did the average bid. Better tracts got more bids and more money.
2. A small percent of the leases produced most of the production—an evidence once again of lognormality of reserve size.
3. More bids and larger amounts of bonus were associated with the highly productive leases. This correlation shows that, in general, industry was able to select the best leases—as evidenced by more bids and more money per tract.
4. Since the ratio of production to bonus declined for leases with large bonus and a higher number of bidders, they infer that winning may not be the sole key to profits. The model of Capen, et al., shows why this occurs, since the winner is the one who most probably overestimated the reserve size.
5. More drilling occurred on the high bonus leases and it started earlier—thus the tendency was for these leases to be productive sooner.

Other Models

Models on competitive bidding have been around for decades. One early reference of two decades ago is by Friedman[6] published in 1956. Its effectiveness was based on previous bidding patterns of competitors. It was designed for single or multiple bidding situations involving contracts.

Of particular interest is a model proposed and tested by Hanssmann.[7] In his book, "Operations Research Techniques for Capital Investment," he writes of an experimental model tested in a sale of mineral leases. He credits Friedman, and Rivett (1959) with similar methodology.

Hanssmann's model employed the following steps:

1. Each prospect is assigned a point score supposedly reflecting value. This assignment was done in consultation with geologists and management.
2. Merging prospects from two prior sales with the upcoming sale all were grouped into four classes A, B, C and D, in order of decreasing value.

3. Industrywide bids from the prior sales established the relative monetary rank of the four groups as 16, 4, 2 and 1. (Hanssmann's ranking was based on several ratios; of particular interest was the ratio of the company's bid in coalitions—they were higher. This situation happens in oil and gas lease bidding and in all competitive bidding. Joint bids are larger than single bids. Why—the risks for the individual companies are lowered! Speitzler[8] found the same relationship. He stated that "group consensus tends to be more risk-taking than the average of individual members of a group").

4. A probability distribution was compiled for the highest competing bid from prior sales for each value group. Lognormal distributions resulted!

5. Final limits were set by management on an objective bidding level of $2.875 million, and a spending level of $1.3 million. Furthermore, no bids were to be placed on A prospects.

6. Hanssmann then combines an estimate of expected value with the distribution made from the highest competitive bid described in 4 above. He is looking for the bid which produces the maximum expected value.

7. From this combination he develops the marginal gain in expected value and searches for the bid, for a prospect, which produces the highest marginal gain for an increment of bid.

8. Final bids are selected for prospects which optimize the expected value at the appropriate bid, using the marginal gain concept.

Hanssmann had a chance to compare his bids with the actual bids at the sale. The company prepared a set of "conventional" bids and used them for the sale. His bids from the study group were not submitted. Thus, a post-sale comparison could be made.

When the results were in, several interesting observations could be made.

- The company decided to bid in coalition with several other companies.
- It did bid on an A prospect.
- More money was spent than the original instructions to Hanssmann's study group.

Hanssmann's Conclusions

Despite these drastic changes in ground rules, Hanssmann makes a comparison of the two systems and concludes:

1. His "scientific" system acquired more acreage at less money than the conventional system. Thus, if the original goals had remained the same, more tracts would have been acquired by his system.
2. His value system acquired more of the value points assigned by the geologists than did the conventional system. This was true even though the A prospect was assigned a high value.
3. The scientific system would have acquired acreage at one-half that paid by the conventional system bidding in coalition.

Hanssmann's model allowed the opportunity of a comparison of the two systems; but in addition, as with all models, it allows the ability to predict the consequences of alternate courses of action. Because his system produced better results, Hanssmann warns against last minute unexpected changes in objectives and strategy. These, he feels, carry the possibility for grave consequences.

The Importance of Competition

Dougherty and Nozaki[9] investigated the significance of competition using a model similar to that of Capen, et al. They found that the optimum bid was a function of three factors.

1. The number of competitors
2. The aggressiveness of the bidding
3. The quality of your own estimate compared to competition.

They conclude further that you should be conservative in bidding when the uncertainty of your own estimate of value increases; i.e., when the risks are greater, the bids should reflect that risk. Dougherty and Nozaki ranked knowing your competitors as of equal value to knowing how well your own estimate stacks up.

Five Important Factors

Hubbell and Deroven[10] used a commercial economic model to test the sensitivity of several factors to rate of return. Their model's acronym was GUESS and is a product of Scientific Software Corp. The acronym stands for General Uncertainty Economic Simulation System. The five factors related to rate of return (ROR) were

- royalty
  - depletion allowance

- crude oil price
- bonus
- reserves

Their work adequately tests a successful prospect for sensitivity to the five factors; however, the element of risk—the chance of complete failure—is not included in their work. As such, their ROR calculations are high and would be sharply reduced by true risk considerations. In addition, their bonus value, $125,000, is unrealistically low for a typical lease in the outer continental shelf of the United States. Nevertheless, the paper does outline methods of testing sensitivity of various parameters.

The computer program they used, GUESS, also has capabilities for entering your variables as distributions to be summed to a final composite distribution by Monte Carlo simulation. Such programs are no better than their input but do allow a view of the complete spectrum of answers possible from your input—and that is important.

REVIEW

From published data about the competition for oil and gas leases in sealed bid sales, we can learn most of the essentials for better competitive bidding. The pertinent points from the literature are:

1. In a sale with many bidders your expectations could possibly be achieved by your average bid; but you will not win on your average bid. You will win when your reserve estimate is overstated relative to the next bidder.
2. Frequently the winning bid is disproportionately higher than the next highest bid, a reflection of lognormality.
3. Because of point 2, there is some room for variations in minimum rate of return, and still make a winning bid.
4. Estimate of reserve size and risk are still the most critical factors in the bid evaluation.
5. However, the roll of competition assumes almost equal importance when the number of bidders is high.

BIBLIOGRAPHY

1. Stark, Robert M., "Competitive Bidding: A Comprehensive Bibliography," Operations Research, Mar-Apr 1971, pgs. 484-490.
2. Crawford, Paul B., "Texas Offshore Bidding Patterns," Journal of Petroleum Technology, March, 1970, pp. 283-289.

3. Capen, E. C., et al, "Competitive Bidding in High Risk Situations," Journal of Petroleum Technology, June, 1971, pgs. 641–653.
4. Megill, R. E., "An Introduction to Exploration Economics," The Petroleum Publ. Co., Tulsa, Oklahoma, 1971.
5. Lohrenz, John and Oden, Hillary A., "Bidding and Production Relationships for Federal OCS Leases . . .," paper given before the 48th annual SPE meeting, Las Vagas, Nev., Sept.–Oct., 1973.
6. Friedman, Lawrence, "A Competitive Bidding Strategy," Operations Research 4, 1956, pgs. 104–112.
7. Hanssmann, Fred, "Operations Research Techniques for Capital Investment," John Wiley & Sons, Inc., N.Y., 1968, pp. 107–126.
8. Speitzler, Carl S., "The Development of a Corporate Risk Policy for Capital Investment Decisions," IEEE Transactions on Systems Science and Cybernetics, Vol. SSC-4, No. 3, Sept., 1968.
9. Dougherty, E. L. and Nozaki, M., "Determining Optimum Bid Fraction," Journal of Petroleum Technology, March, 1975, pp. 349–356.
10. Hubbell, Robert O., and Deroven, Gordan A., "How Five Major Bid Factors Affect Potential Lease Profit," World Oil, July, 1975, pp. 67–69.

16 The Fundamentals of Risk Analysis

This final chapter summarizes the principles involved in the analysis of risk. We will be summing up the fundamentals, commenting on some problems inherent in all risk analysis and reviewing the steps in setting up a "solution" to a problem requiring an analysis of risk.

First a review of the fundamental concepts necessary for the best analysis of risk.

The Fundamentals

In the search for new oil and gas fields, risk analysis takes the judgments of explorationists and engineers and translates them into the language of probability. Risk analysis, thus, helps a manager make reasonable decisions where the key parameters have considerable uncertainty.

The mathematical models used to synthesize risk analysis employ Monte Carlo simulation. Occasionally, we find the mistaken conclusion that Monte Carlo simulation is risk analysis. The word simulation here is the clue. The judgments we input as probability distributions can be summed to a final distribution. That distribution simulates our concept of the risks involved. It is only as good as our input. It takes into account only the variables submitted.

1. The first fundamental of risk analysis, then, is this: You must know enough about your proposed analysis to isolate the key variables. An obvious or inadvertent omission will result in poor output. It may be so poor, in fact, that you have not analyzed the real risk at all! Furthermore, we constantly face the danger, when facts are few, that our concept of how a parameter may vary is extremely limited.
2. After isolating the key variables then you must find a means to quantify these variables. Unfortunately, quantification faces problems.

Objective vs Subjective Probability

Type of Drilling	Type of Probability	
	Objective	Subjective
Development	X	
Outpost	X	X
Wildcat—Mature Trend	X	X
Wildcat—Frontier		X

FIG. 16.1

For example, Fig. 16.1 illustrates the varying amounts of data affecting drilling decisions. The illustration is based on the premise that for objective probability we have some real world empirical experience upon which to base probabilities or quantification. Subjective probability stems from essentially no data.

On the first line of Fig. 16.1, development drilling, we have the most information. The statistical base here gives a reasonable view of the future and some assurance of real world probabilities. Development drilling typifies objective probability.

Outpost drilling contains less data, and, therefore, requires some subjective judgments. Wildcat drilling in a mature trend has a reasonable data base, but more subjective judgments. In a frontier we may have extremely limited data. Some theorists would say that in wildcat drilling we have no data base from which to estimate probabilities. Others would say your opinions— technical judgments or "gut" feelings—are a form of subjective probability.

In prior chapters, we used single number estimates of probability (geologic success estimates) and triangular distributions to quantify key variables. The choice depends upon the mathematical model used and how your data are input. Sometimes data are so few that the geologist faces a real dilemma in trying to describe his uncertainty.

On Fig. 16.2 we see the situation faced by many explorationists. With few facts, many explorationists feel comfortable only with words—such as poor, fair, good or excellent. The theorist says numerical expression allows the discipline of math to describe dissimilar opportunities with greater consistency and adequacy. But when the facts are few, our opinions range widely. We should expect it to be so. Even quantifying the words poor, fair, good

Expressing Uncertainty—Ranking a Variable—

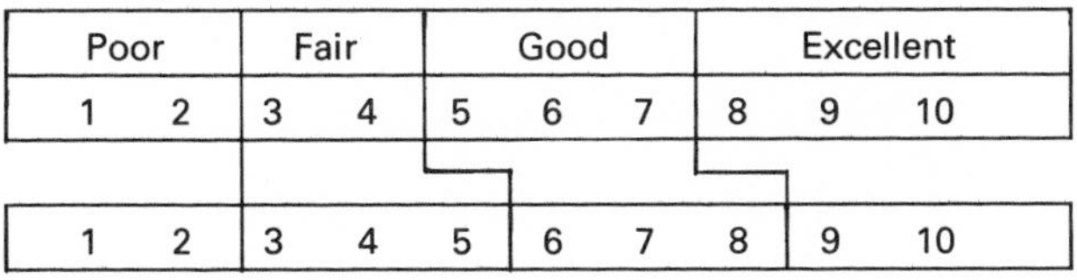

Poor		Fair		Good			Excellent		
1	2	3	4	5	6	7	8	9	10

1	2	3	4	5	6	7	8	9	10

FIG. 16.2

and excellent can be done in more than one way, as shown in Fig. 16.2.

Despite the fears associated with quantification the second fundamental is absolutely necessary. Use your best judgment and find a satisfactory method of quantifying your uncertainty.

3. Another fundamental consideration in risk analysis involves your basic view of uncertainty. Most parameters have up-side and down-side possibilities in addition to a most probable value. Your view of a parameter should reflect both the up-side as well as the down-side. A triangular distribution does reflect these values, one reason for its popularity.

It is in time related variables that the possibility exists for a limited view of the variable. Fig. 16.3 illustrates this problem. Here are two views of uncertainty; on the left is what some have called the cone of uncertainty. It reflects, in forecast form, both positive and negative views of the future. On the right is what one of my associates calls the horse-tail view of uncertainty. Only negative tomorrows are viewed here. This view asks only what alternatives can produce bad results. White, et al.,[1] have shown that even extrapolations related to time can be constructed to give the up-side and down-side extrapolations on a basis related to history.

In quantifying any parameter consider the good outcomes as well as the possibly bad outcomes.

4. A fourth fundamental of risk analysis involves understanding your model. Even if your model is a simple equation we should comprehend its affect on your input. This admonition was stated in Chapters Ten and Eleven regarding triangular distributions; but it applies equally to the model itself.

If you don't have or can't understand a mathematical consultant, you can test your model independently. Substitute a range of

Two Views of Uncertainty

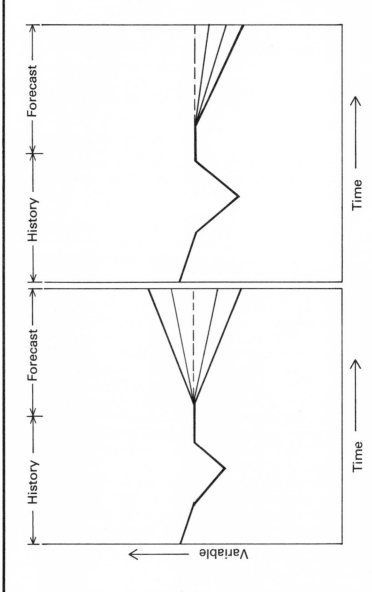

FIG. 16.3

cases and illustrations until you understand the effect on the output of the changing conditions of your input.

One of the dangers of the computer age is that users assume the programs are all written properly. Often a program may have some flaw not detected by the user. A hand calculated case, obviously simple as a matter of convenience, will often expose a false concept or equation in your model.

5. A most important fundamental of risk analysis was brought out in Chapter 14. In the discussion on the Delphi Method one fundamental was stressed—put your estimates of probability into your key variables BEFORE simulation in your model. Don't "massage" the final answer (except in the case next discussed) for probability adjustments—go back to the basic ingredients. There you have a better grasp of the limits and your input should reflect your judgments at that point. An exception to this occurs in our next fundamental.

6. The scientist, analyst, forecaster must constantly strive for credibility. He achieves what credibility is possible—and complete credibility is never possible for the unknown—by tying to reality.

A fundamental of risk analysis, therefore, is to search for reality checks. These can be found by comparison to similar but known situations, by correlation, or by checks of limits set by reality. Search then for means to support credibility.

If you hold your answers up to the light of reality and you don't like the result, what then? Go back to your input. Answers in risk analysis are only as sound as their weakest assumption. Re-examine your input and see what factor produced the questionable answer.

Some examples of answers lacking credibility or at least rating a recheck of the input are:

a. The mean field size estimated is larger than any field yet discovered in the trend.
b. Your composite geologic success factor for a number of prospects is greater than the trend average.
c. Your assessment of a basin produces recoveries higher than comparable basins in the world.
d. Your field sizes produce well sizes (barrels or Bcf per well) larger than previously known in the trend.

Many other examples are possible. Each of the preceding four examples show cause for rechecking even though the analysis may be showing you the right answer. You could have the best prospect in the trend. Your basin could be better than similar basins around the world; your sand thickness could be above the average for the trend. Nevertheless, for credibility's sake you need a ready explanation for why you think so; and you may have to go back to your beginning assumptions to substantiate your case.

7. As a final fundamental, express the uncertainty of your solution in the form of a cumulative probability distribution. Such a distribution best expresses the range of uncertainty in a complex problem or projection.

In addition:

a. It demonstrates the fallacy of a single answer.

b. It illustrates all foreseeable answers based on your estimates of uncertainty.

c. It provides a mean value and shows the most frequently occurring values at its steepest slope.

d. It reminds us again that we do our managers a disservice when we show only a single answer to a problem, forecast, or projection which has great uncertainty.

The Never Ending Problems Associated With Risk Analysis

After you have done your very best in setting up an analysis of risk, have made reality checks and have satisfied yourself that you have considered all of the right variables, you will still have problems. Many of them are never-ending ones.

Yearn for Certainty

One example is management's natural yearn for certainty. In a large corporation much time is spent trying to reduce risk as much as possible. Sometimes this yearn for certainty makes us victims of the search for a single answer. The worst thing an analyst can do for a manager is to obscure the line between what is known and what is unknown.

Cut-It-In-Half

Another problem in risk analysis is associated with the "cut-it-in-half" manager. We've all met one. Maybe you've been one! The idea is that if it still looks good with half the reserves it's safe. Let's face one fact, however. One misses some good deals with such a pessimistic view of the future. For every "cut-it-in-half" manager we may need a "two-times" manager. The future can be both positive and negative. If every analysis faces a 50% cut a lot of time and money may be wasted on risk analysis. You don't really look at the risks when you "cut-it-in-half"; and you are ignoring all prior risking. It is a true example of "double-dipping" your risking process.

The reason this problem will never completely go away is that in most corporations the penalty for errors of commission is several fold greater than penalties for errors of omission.

Over-Optimism

One problem in risk analysis particularly affects explorationists. Our profession, because of its high chances for failure, puts a premium on optimism and on selling geologic ideas. Sometimes this aspect gets out of hand. It results in over-sell and we inadvertantly blind ourselves to the flaws in our analysis. Optimism *is* important, but so is realism!

Experience

Strangely, one problem is experience. Judgment and experience are important. They are just as important today as at any period in history. ("Judgment comes from experience and experience from bad judgment.") However, in a rapidly changing world, experience can be a handicap. Each of us needs to have the humility to remember the new ideas we rejected—ideas which later proved successful. It's no crime to be limited by our experiences. It's a shame, however, to have our experiences result in a closed mind.

The All-Important Outcome. A big problem in risk analysis and decision making under vast uncertainty is separating the right decision from the outcome. Uncertainty means the outcome can have wide variance. Often a good tactical decision is made, but the outcome is poor. Unfortunately, the world of reality tends to judge the quality of a decision only by the outcome. Often "Monday morning quarterbacks" don't have all the facts or choose to ignore a few.

Good outcomes can come from bad decisions and poor outcomes can result from good decisions. When a bad decision is obscured by serendipity or a fortuitous outcome, risk analysis takes a ribbing. However, in a long-term game of high risks you will go broke without an intelligent, consistent way to handle risk.

A Review of Steps in Risk Analysis

Our previous section on the fundamentals of risk analysis is almost in itself a step-by-step approach to an analysis of risk. Nevertheless, a brief review will reiterate the significant steps in an analysis of risk.

1. Gather your data. A normal start in any analysis, this step is no less important for an analysis of risk. One of the initial things you want to discover is how much is not known.
2. Isolate the key variables. Past experience may have to be a key. On the other hand, you may not know what some of the key variables are until you have gone through the problem and run some sensitivity cases.
3. Quantify the key variables. Triangular distributions are recommended for uncertain variables with many possible answers. For other variables, consider the positive as well as the negative to avoid over-risking. For exploratory problems remember the important significance of lognormality.
4. Be sure that your concepts of uncertainty are put in at the variable level, not at the final answer level.
5. Enter your input into whatever model you use.
6. Check the answer for reality. Protect your credibility.
7. Express the final answer in the form of a cumulative frequency distribution. This approach will not let you fall victim to the single answer syndrome. In a very uncertain investment (one with many possible outcomes) the probability of a single value being the answer is near if not zero.

Finally we need to remember one last concept. Every complex business decision has to be made in light of the relevant facts which are known—but also in recognition of what is unknown and possibly unknowable.

Risk analysis provides insights when the unknown is a major factor. It helps sort reasonable approaches out from many possible approaches.

Current concepts of risk analysis, although far from perfect, do provide better insights about investments with great uncertainty.

The Risk of the Unique Event

Exploration is a process which commits company funds to an unknown future. The unknowns involve not only geologic uncertainty but a number of critical economic factors such as price, cost, inflation, and possible changes in tax laws. Exploratory funds, therefore, are committed to a sum of unknowable expectations—not to facts.

It is this challenge which lends the excitement to the search for oil and gas fields. It is this element which constantly reminds the explorationist that risk is the essence of his business—that risk taking represents the basic and constantly underlying principle of his work.

All business enterprises face risks and unknowns. The important difference for the explorationist is the number of times he is faced with a unique, unrepeatable investment opportunity for which there is no prior experience or statistical probability to guide his decision.

The explorationist understands that he can never eliminate risk from his work. But he also knows that the best results from his efforts will be that management will have taken the right risks.

BIBLIOGRAPHY

1. White, D. A. et. al., "Assessing Regional Oil & Gas Potential," from Methods of Estimating the Volume of Undiscovered Oil and Gas Reserves, AAPG Studies in Geology No. 1, 1975, pg. 143.

Epilogue

If the goals set forth in the beginning of this book have been achieved, then your further reading in the arena of risk analysis will be easier. You will have mastered and gained new insights into some of the fundamentals of the analysis of risk. You are the judge.

There are other areas of study in the field much more complex yet deserving of study. Utility theory, Bayesian concepts, and more complex statistical applications would add to your knowledge. This introduction should make these additional readings easier to absorb and use.

The many ideas and theories relating to the analysis of risk are being challenged today and have always been challenged.[1,2] Until more professionals understand both the benefits and limitations of risk analysis, managers will go on making difficult decisions without the help of a better understanding of the dimensions of their decision. The real challenge of risk analysis is to the teacher who must find simple, semimathematical means of explaining difficult concepts. He must also, if given the opportunity, show that his methods can produce results not only better than other methods, but at less cost. Once this is accomplished he will find a ready audience for his work.

BIBLIOGRAPHY

1. Hall, William K., "Why Risk Analysis Isn't Working," Long Range Planning, December 1975, page 25.
2. Carter, E. S., "What are the Risks of Risk Analysis," Harvard Business Review, 1972, pages 72-82.

Appendix A

Triangular Distribution

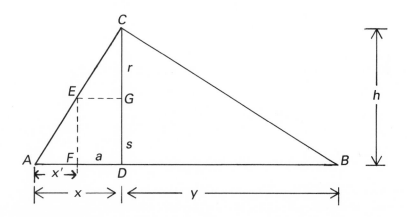

Given: $h = r + s$

$a = x - x'$

This appendix shows that the area of any part of a triangle can be expressed as a percent of the total triangle in terms of values of the base line alone.

Before proceeding further, however, please refer to the Reference Triangle which is the last page of this appendix. Its purpose is to familiarize you with the basic segments of the triangle which will be used in the appendix. Below the triangle you will find definitions of the terms. After reviewing this page, you are ready to proceed.

Part I

Our aim in these relationships will be to find ways of defining area entirely as functions of x and y. To begin with, we will make use of the obvious fact that the area of $\triangle AFE$ is equal to the area of $\triangle ADC$ *minus* the area of the polygon *FDCE*.

Setting this thought up in the form of an equation (remember that the area of a $\triangle$ is one-half the base times the height) we have:

$$\frac{sx'}{2} \text{ (the area of } \triangle AFE) = \frac{xh}{2} \text{ (area of } \triangle ADC) - sa \text{ (area of polygon}$$

$$FDGE) - \frac{ra}{2} \text{ (area of } \triangle EGC)$$

which without the words reduces to $\dfrac{sx'}{2} = \dfrac{xh}{2} - sa - \dfrac{ra}{2}$; multiplying both sides of the equation by 2 we have:

$$sx' = xh - 2sa - ra = xh - sa - sa - ra$$

$$sx' = xh - ra - sa - sa = xh - (r + s)a - sa$$

We know that $(r + s) = h$, so we can substitute and have:

$$sx' = hx - ha - sa$$

We now substitute for a which, as we said, equals $(x - x')$.

$$sx' = hx - h(x - x') - s(x - x') = hx - hx + hx' - sx + sx'$$

The hx values cancel and we have:

$$sx' = hx' - sx + sx'$$

which reduces to

$$hx' = sx$$

We have covered a lot of "equation" territory for a simple relationship. If you are interested, you can check our simple relationship by constructing a triangle on graph paper with rectangular coordinates and you will see that hx' does always equal sx. Our use of this simple equation will be to substitute for values of s and from the equation we know that

$$s = \frac{hx'}{x}$$

Part II

Now we can go back to our main purpose—to show that a part of the total area can be expressed as a function of the total area in terms of the base line alone. As we will see later, we have to express this idea in two segments. One segment deals with the areas left of the most likely value, the other segment with areas to the right. We shall deal first with the area to the left of the most likely value.

This time, rather than using "a" we shall begin with $(x - x')$ as the value of a. Our ratio of any triangular area to the left of the most likely value (M.L.) becomes:

$$\frac{\dfrac{xh}{2} - s(x - x') - \dfrac{r}{2}(x - x')}{\dfrac{h(x + y)}{2}} = \frac{\text{area of any } \triangle \text{ left of M.L.}}{\text{area total } \triangle}$$

Multiplying the top and bottom of the fraction by 2, we have:

$$\frac{hx - 2sx + 2sx' - rx + rx'}{h(x + y)}$$

Rearranging to take advantage of the relationship

$$r + s = h$$

we have:

$$\frac{hx - (r + s)x + (r + s)x' - sx + sx'}{h(x + y)} = \frac{hx - hx + hx' - sx + sx'}{h(x + y)}$$

The hx values cancel out and we have:

$$\frac{hx' - sx + sx'}{h(x + y)}$$

but, $hx' = sx$ and substituting for sx, the hx' values cancel out and our equation becomes:

$$\frac{sx'}{h(x + y)}$$

From the previous section we developed the relationship that $s = \dfrac{hx'}{x}$; substituting for the value of s we have:

$$\frac{\dfrac{hx'x'}{x}}{h(x + y)} \quad \text{which equals} \quad \frac{\dfrac{(x')^2}{x}}{(x + y)} \quad \text{(the } h\text{'s cancel)}.$$

We can show our ratio now in an equation form as follows:

$$\mathop{CF}_{\substack{x'=x \\ x'=0}} = \frac{\dfrac{(x')^2}{x}}{(x + y)}$$

which is read: the cumulative frequency for values of x' from 0 to x equals x' squared, divided by x, and this quantity then divided by the sum of $(x + y)$.

Newendorp[1] defines $\dfrac{x}{x + y}$ as "m" and x' as $x(x + y)$ or $x = \dfrac{x'}{x + y}$. His form for the equation is $\dfrac{x^2}{m}$ which in our nomenclature would be shown as:

$\dfrac{\left(\dfrac{x'}{x + y}\right)^2}{\dfrac{x}{(x + y)}}$. It can be reduced to our equation of $\dfrac{(x')^2}{\dfrac{x}{x + y}}$. Going back to our

equation, you can see that when $x' = x$, the ratio reduces to a simple $\dfrac{x}{x + y}$.

So, in triangular distributions at the most likely value the cumulative frequency of values equal to or less than $(\leq) x$ will be $\dfrac{x}{x + y}$.

Part III

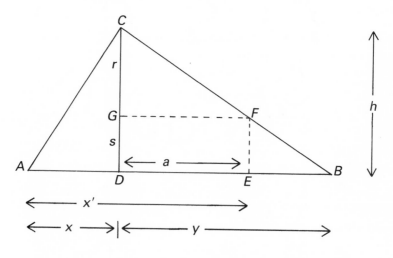

$$a = x' - x$$
$$h = r + s$$

We will use the same approach for the right side of the most likely value as for the left. We can sum up the areas as follows:

Area EBF = Area ABC − Area ADC − Area $DEFG$ − Area GFC

$$\frac{s\left[(x + y) - x'\right]}{2} = \frac{h(x + y)}{2} - \frac{xh}{2} - sa - \frac{ra}{2}$$

Multiplying both sides of the equation by 2 we have:

$$sx + sy - sx' = hx + hy - hx - 2sa - ra$$

The hx's cancel and we substitute for a which equals $(x' - x)$

$$sx + sy - sx' = hy - 2s(x' - x) - r(x' - x)$$
$$= hy - sx' - sx' + sx + sx - rx' + rx$$

Combining $sx + sx'$ values on the right side we have:

$$sy = hy - rx' - sx' + rx + sx = hy - (r + s)x' + (r + s)x$$

but $(r + s) = h$ so, $sy = hy - hx' + hx$ and our beginning relationship becomes

$$sy = h(y - x' + x).$$

We will use this relationship to substitute, later, for s. You can prove this mathematical statement by drawing a triangle to some easy scale and inserting the proper numbers in the equation.

Part IV

We now return to our area ratios. For the right side of x we will use the ratio of the little triangle EBF as related to the large triangle ABC.

Our ratio is:

$$\frac{\dfrac{s(x + y - x')}{2}}{\dfrac{h(x + y)}{2}} = \frac{\text{Area of the small } \triangle}{\text{Area of the whole } \triangle}$$

Multiplying both numerator and denominator by 2 we have:

$$\frac{s(y - x' + x)}{h(x + y)}$$

Now we can substitute for the value of s derived above and we have:

$$\frac{\dfrac{\cancel{h}(y - x' + x)}{y}(y - x' + x)}{\cancel{h}(x + y)}$$

The h's cancel; multiplying top and bottom by y we have:

$$\frac{(y - x' + x)(y - x' + x)}{y(x + y)}$$

or:

$$\frac{y^2 - yx' + xy - yx' - x'^2 - xx' + xy - xx' + x^2}{y(x + y)},$$

Rearranging we have:

$$\frac{x^2 + 2xy + y^2 - 2yx' - 2xx' + x'^2}{y(x + y)}$$

Factoring we have:

$$\frac{(x + y)^2 - 2x'(x + y) + x'^2}{y(x + y)} = \frac{[(x + y) - x']^2}{y(x + y)}$$

Note for the special case of $x = x'$ this ratio reduces to $\frac{y}{x + y}$. We shall need this relationship later.

Returning to the form of our ratio as: $\dfrac{(x + y)^2 - 2x'(x + y) + x'^2}{y(x + y)}$ and dividing

by $(x + y)^2$ we then have: $\dfrac{1 - \dfrac{2x'}{x + y} - \dfrac{x'^2}{(x + y)^2}}{\dfrac{y}{(x + y)}}$; however, by definition we

know that $\dfrac{x}{x + y} + \dfrac{y}{x + y} = 1.0$. Therefore $\dfrac{y}{(x + y)} = 1 - \dfrac{x}{(x + y)}$; our ratio now becomes:

$$\frac{\left[1 - \dfrac{x'}{(x + y)}\right]^2}{1 - \dfrac{x}{x + y}}$$

This is the ratio of the very small triangle to the large triangle. To get the proper value for cumulative frequency we must subtract this value from one. By doing this we get the ratio of the area up to x' which is what we want, not the area of the small triangle. So our final expression becomes:

$$\begin{array}{c} x'=(x+y) \\ CF \\ x'=x \end{array} = 1 - \frac{\left[1 - \dfrac{x'}{(x + y)}\right]^2}{1 - \dfrac{x}{(x + y)}}$$

We can prove the logic of this last step several ways. Consider only the ratio (not the ratio subtracted from 1.0). What happens as x' approaches $(x + y)$ the maximum? The ratio goes to zero—as it should because the area of the small triangle goes to zero.

Consider another way. What happens to the ratio when x' becomes x? Under this condition the ratio becomes $\dfrac{y}{x + y}$ and the small triangle has enlarged to become the exact complement of $\dfrac{x}{x + y}$. This latter value was the ratio for the small triangle starting from the left of x. This proof satisfies our definition $\dfrac{x}{x + y} + \dfrac{y}{x + y} = 1.0$ and says that the area to the left of the M.L. value plus the area to the right of the M.L. value equals one. This was our beginning assumption.

Newendorp [1] defines $\dfrac{x'}{x + y}$ as x and $\dfrac{x}{(x + y)}$ as m, so his ratio is shown as:

$$1 - \frac{(1 - x)^2}{1 - m}$$

Summary

Using triangular distributions the base line values, alone, can be used to convert the distribution to cumulative frequency values. Referring to the reference triangle, we have these definitions:

x = the most likely value minus the minimum
y = the maximum value minus the most likely
x' = represents any and all values along the base line (minus the minimum value)

Our final equations are:

1. For values of x' equal to or less than ($\leq$) x

$$\underset{x'=0}{\overset{x'=x}{C F}} = \frac{\dfrac{(x')^2}{x}}{(x + y)}$$

2. For values of x' equal to or greater than ($\geq$) x

$$\underset{x'=x}{\overset{x'=(x+y)}{C F}} = 1 - \frac{\left[1 - \dfrac{x'}{(x + y)}\right]^2}{1 - \dfrac{x}{(x + y)}}$$

Reference Triangle

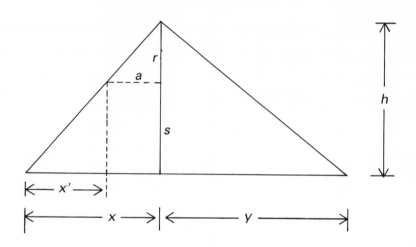

x = most likely minus the minimum
y = maximum value minus the most likely
x' = any value of x from the minimum to the maximum
h = height of triangle
$r + s$ are segments of h, such that $(r + s) = h$
a is a segment of $x + y$
   —to the left of x it is defined as $x - x'$
   —to the right of x it is defined as $x' - x$
$x + y$ = entire range of values of x' from the minimum to the maximum

Remember that the area of any triangle is equal to one-half the base times the height. Thus for the large triangle the area is:

$$\frac{1}{2}(x + y)\,h$$

BIBLIOGRAPHY

1. Newendorp, Paul D., "Decision Analysis for Petroleum Exploration," Petroleum Publishing Co., Tulsa, Okla., 1975, pg. 274.

Appendix B

Shortcuts to Calculating the Mean

Given a cumulative-frequency distribution, it is often desirable to know the mean of the distribution. Several methods of estimating the mean and standard deviation are available from the field of order statistics.[1]

The USGS uses the approximate method of adding the values at the 5 and 95 percentile values to the modal value and dividing by three. They refer to this as a "statistical mean."[2] The difficulty with this method is that you must know, calculate, or have given the modal value.

A much simpler method was discovered by R. I. Swanson.[3] In this method all data can be read from the cumulative-frequency distribution—either the linear plot or the plot on log probability paper. Three steps are involved:

1. Read the parameter values from the 10th, 50th, and 90th percentile intersections.
2. Multiply these values by probabilities of .3, .4, and .3 respectively.
3. Sum the risk-weighted values for the mean.

This method produces a mean which is amazingly close to the mean generated by 5,000 iterations from Monte Carlo simulation. The accuracy is most often within 1% of the real value.

A sample calculation is illustrated as follows:

Percentile	Variable Value	Probability	Risk-weighted Variable Value
10	50	.3	15
50	100	.4	40
90	200	.3	60
		Mean	115

The mean in this illustration is 115—larger than the median (100) indicating a right-skewed distribution.

The Uses

The mean is the only value which can be added directly from one distribution to another. For this reason alone, it is a popular parameter for a distribution. The mean best expresses a single number value, if you must settle on only one value, for a distribution. So, the mean of reserve sizes, profit, discounted profit, etc., can be indicative of the relationship of one distribution to another.

Another practical use involves profit calculations. If you calculate the profit or present-value profit for the 10th, 50th, and 90th percentile prospect sizes you can use the same probabilities to calculate the profit for the mean.

For example:

Percentile	PVP @ 10%	Probability	Risk-Weighted PVP @ 10%
10	100	.3	30
50	200	.4	80
90	400	.3	120
		Mean PVP @ 10%	230

You will find this last shortcut quite valuable in using distributions and relating them to other economic relationships.

BIBLIOGRAPHY

1. Eisenberger, Isidore and Posner, Edward C., "Systematic Statistics used for Data Compression in Space Telemetry," Journal of the American Statistical Association, March 1965, Vol. 60, No. 309, page 97.
2. Miller, Betty, M., et al., "Geological Estimates of Undiscovered Recoverable Oil and Gas Resources in the United States," Geological Survey Circular 725, 1975, page 21.
3. Swanson, R. I., unpublished memorandum, Sept. 1972.

Appendix C

Individual Binomial Probability

$B(x, n, 0.30)$

Values of x

TABLE 1

n	0	1	2	3	4	5	6	7	8	9	10	11	12	13	14	15	16	17	18
2	.49	.42	—																
3	.34	.44	.19	—															
4	.24	.41	.26	.08	—														
5	.17	.36	.31	.13	.03	—													
6	.12	.30	.32	.19	.06	.01	—												
7	.08	.25	.32	.23	.10	.03	—												
8	.06	.20	.30	.25	.14	.05	.01	—											
9	.04	.16	.27	.27	.17	.07	.02	—											
10	.03	.12	.23	.27	.20	.11	.04	.01	—										
11	.02	.10	.20	.26	.22	.13	.06	.02	—										
12	.01	.07	.17	.24	.23	.16	.08	.03	.03	.01	—								
13	.01	.05	.14	.22	.23	.18	.10	.04	.01	—									
14	—	.04	.11	.19	.23	.20	.13	.06	.02	.01	—								
15	—	.03	.10	.17	.22	.21	.15	.08	.03	.01	—								
20	—	.01	.03	.07	.13	.18	.19	.16	.11	.07	.03	.01	—						
25	—	—	.01	.02	.06	.10	.15	.17	.17	.13	.09	.05	.03	.01	—				
30	—	—	—	.01	.02	.05	.08	.12	.15	.16	.14	.11	.07	.04	.02	.01	—		
35	—	—	—	—	.01	.02	.04	.07	.10	.13	.14	.14	.12	.09	.06	.04	.02	.01	—
40	—	—	—	—	—	.01	.02	.03	.06	.08	.11	.13	.14	.13	.10	.07	.05	.03	.02
45	—	—	—	—	—	—	.01	.01	.03	.05	.07	.10	.12	.13	.13	.11	.09	.07	.04

Appendix C

Cumulative Binomial Probability

B(x or more, n, 0.10)

Values of x   TABLE 2

n	0	1	2	3	4	5	6	7	8	9	10	11	12	13	14	15	16	17	18
2		.19	.01	—															
3		.27	.03	—															
4		.34	.05	—															
5		.41	.08	.01	—														
6		.47	.11	.02	—														
7		.52	.15	.03	—														
8		.57	.19	.04	.01	—													
9		.61	.23	.05	.01	—													
10		.65	.26	.07	.01	—													
11		.69	.30	.09	.02	—													
12		.72	.34	.11	.03	—													
13		.75	.38	.13	.03	.01	—												
14		.77	.42	.16	.04	.01	—												
15		.79	.45	.18	.06	.01	—												
20		.88	.61	.32	.13	.04	.01	—											
25		.93	.73	.46	.24	.10	.03	.01	—										
30		.96	.82	.59	.35	.18	.07	.03	.01	—									
35		.97	.88	.69	.47	.27	.13	.06	.02	.01	—								
40		.99	.92	.78	.58	.37	.21	.10	.04	.02	.01	—							
45		1−	.95	.84	.67	.47	.29	.16	.08	.03	.01	—							

Appendix C

Cumulative Binomial Probability

$B(x$ or more, n, 0.20)

n	0	1	2	3	4	5	6	7	8	9	10	11	12	13	14	15	16	17	18
2	.36	.04	—																
3	.49	.10	—																
4	.59	.18	.03	—															
5	.65	.26	.06	—															
6	.74	.34	.10	.02	—														
7	.79	.42	.15	.03	—														
8	.83	.50	.20	.06	.01	—													
9	.87	.56	.26	.09	.02	—													
10	.89	.62	.32	.12	.03	.01	—												
11	.91	.68	.38	.16	.05	.01	—												
12	.93	.73	.44	.21	.07	.02	—												
13	.95	.77	.50	.25	.10	.03	.01	—											
14	.96	.80	.55	.30	.13	.04	.01	—											
15	.96	.83	.60	.35	.16	.06	.02	—											
20	.99	.93	.79	.59	.37	.20	.09	.01	—										
25	1−	.97	.90	.77	.58	.38	.22	.11	.05	.02	.01	—							
30	1	.99	.96	.88	.74	.57	.39	.24	.13	.06	.03	.01	—						
35	1	1	.98	.94	.86	.73	.57	.40	.25	.07	.03	.01	.01	—					
40	1	1	.99	.97	.92	.83	.71	.56	.41	.27	.16	.09	.04	.02	.01	—			
45	1	1	1	.99	.96	.91	.82	.70	.56	.41	.30	.17	.10	.05	.02	.01	—		

Appendix C

Cumulative Binomial Probability

$B(x$ or more, n, 0.30)

Values of x TABLE 4

n	0	1	2	3	4	5	6	7	8	9	10	11	12	13	14	15	16	17	18
2	.51	.09	—																
3	.66	.22	.03	—															
4	.76	.35	.08	.01	—														
5	.83	.47	.16	.03	—														
6	.88	.58	.26	.07	.01	—													
7	.92	.67	.35	.13	.03	—													
8	.94	.74	.45	.19	.06	.01	—												
9	.96	.80	.54	.27	.10	.03	—												
10	.97	.85	.62	.35	.15	.04	.01	—											
11	.98	.89	.69	.43	.21	.08	.02	—											
12	.99	.91	.75	.51	.28	.12	.04	.01	—										
13	.99	.94	.80	.58	.35	.17	.06	.02	—										
14	.99	.95	.84	.64	.42	.22	.09	.03	.01	—									
15	1	.96	.87	.70	.48	.28	.13	.05	.02	—									
20	1	.99	.96	.89	.76	.58	.39	.23	.11	.05	.02	.01	—						
25	1	1	.99	.97	.91	.81	.66	.49	.32	.19	.10	.04	.02	.01	—				
30	1	1	1	.99	.97	.92	.84	.72	.57	.41	.27	.16	.08	.04	.02	.01	—		
35	1	1	1	1	.99	.97	.94	.87	.77	.64	.49	.35	.23	.14	.07	.04	.02	.01	
40	1	1	1	1	1	.99	.98	.94	.89	.80	.69	.56	.42	.30	.19	.12	.06	.03	
45	1	1	1	1	1	1	.99	.98	.95	.91	.84	.74	.62	.49	.37	.25	.16	.10	

Appendix C

Probability of All Dry Holes

TABLE 5

n	p = .1	p = .2	p = .3
2	.81	.64	.49
3	.73	.51	.34
4	.66	.41	.24
5	.59	.35	.17
6	.53	.26	.12
7	.48	.21	.08
8	.43	.17	.06
9	.39	.13	.04
10	.35	.11	.03
15	.21	.04	
20	.12	.01	
25	.07		
30	.04		
35	.03		
40	.01		

(x = 0)

Index